SDGs and Textiles

Series Editor

Hafeezullah Memon, International Institute of Silk, Zhejiang Sci-Tech University, Hangzhou, China

The book series "SDGs and Textiles" addresses the strategies to achieve sustainable development goals (SDGs) in the present, past, and future. It presents books about the present and future policies of textile ministries of different countries, and books related to sustainability education around different parts of the world in the textile sector. Moreover, it would welcome the conference proceeding related to SDGs and Textiles. The series would cover books comparing the sustainability and SDGs of different institutions and countries. The individual book volumes in the series are thematic. The goal of each book is to give readers a comprehensive overview of a different area of sustainability in the textile sector. As a collection, the series provides valuable resources to a broad audience in academia, the research community, industry, and anyone looking to expand their knowledge of SDGs and Textiles.

Textiles and life are together – life cannot be separated from textiles as it is the most important need for human beings after food. In 2015, the United Nations General Assembly proposed 17 interlinked global goals to be achieved by 2030. Since then, academia and industry have paid much attention to achieving these goals. Textile found its close relation with almost all of these 17 goals.

SDG 1 - No Poverty: Poverty would never be overcome by a charity only; it is essential to develop people's skills to have a better and wealthy life. Thus, the textile can be considered an excellent discipline to achieve this goal by creating jobs and small and medium businesses.

SDG 2 - Zero Hunger: Through the effective utilization of advanced application of Agrotech Textiles, it is possible to have higher crop yields and save crops from rough weather, unexpected rains, floods, insects, etc.; thus, geotextiles play an essential in achieving this goal of sustainable development.

SDG 3 - Good Health & well-being: There has been much health consciousness after Covid19, and medical textiles assist in getting good health and well-being.

SDG 4 - Learning & Education: Textile or fashion has remained a significant discipline for societies for ages, and there has always remained much to explore in this field. Textile-related universities may play a vital role by offering free access to their education resources, training and spreading information among the locals.

SDG 5 - Gender Equality: The textile sector is one of the industrial sectors that accepted gender equality long ago; in particular, the garment sector has more females than males. Thus, the textile sector has been doing gender equality. Moreover, there has been a recent trend for Gender Neutral Clothing, which need worth studying and may further assist gender equality.

SDG 6 - Clean Water & Sanitation: Textiles could be achieved through filtration, and of course, textile is one of the critical materials for filtration.

SDG 7 - Affordable & Clean Energy: With the recent advancement in material science and engineering, the textile sector has come on the front for, not only by using this clean energy during textile production but also by assisting the production of this clean energy, either in the form of wind turbines blades made of textile composites or by energy harvesting from T-Shirts, etc.

SDG 8 - Decent Work: Recently, there has been much attention that the textile workers are not paid well, labor rights are not cared about, etc.

SDG 9 - Industry and innovation: Textile Industry always follows innovation; the textile companies that do not chase innovation cannot survive in the market.

SDG 10 - Reduced Inequalities: Getting better life and well-being would help reduce inequalities in the textile industry.

SDG 11 - Sustainable Cities: Sustainable Textile Cities through Buildtech and transport textiles.

SDG 12 - Consumption and Production: Textile and garment consumption and production all come under.

SDG 13 - Climate Action: Oekotech or Ecotech Textile, waste management of textiles are upfront to achieve this goal of sustainable development.

SDG 14 - Life Below Water: Mitigating microfiber waste in rivers and oceans may come under the context of it. There has been much attention on this subject after passing the bill at the parliament level of the UK.

SDG 15 - Life on Land: Geotech or Geotextiles studies life on land.

SDG 16 - Peace, Justice, and Strong Institutions: Protective textiles are doing their best to achieve peace, justice, and strong institutions.

SDG 17 - Partnerships for the Goals: The application of textiles to achieve sustainable development goals is only an example. In all textiles sectors, combined efforts of all the goals are essential to achieve true sustainability.

José Fernando Gallego-Nicholls ·
Agustín Carrilero-Castillo
Editors

Textile Industry and the SDGs

Exploring Synergies for a Better Future

Springer

Editors
José Fernando Gallego-Nicholls
ESIC Business & Marketing School
Valencia, Spain

Agustín Carrilero-Castillo
ESIC Business & Marketing School
Valencia, Spain

ISSN 2948-1236 ISSN 2948-1244 (electronic)
SDGs and Textiles
ISBN 978-981-97-5072-6 ISBN 978-981-97-5073-3 (eBook)
https://doi.org/10.1007/978-981-97-5073-3

This Springer imprint is published by the registered company Springer Nature Singapore Pte Ltd.
The registered company address is: 152 Beach Road, #21-01/04 Gateway East, Singapore 189721, Singapore

Preface

The vast impact of the textile industry in modern society is undeniable. Around 70 million jobs, out of which two-thirds are based in Asia, is the size of its workforce. This is about 1% of the entire global human population. Now, regardless of our livelihood, today everyone needs to wear clothes and shoes. In short, the textile industry affects us all. This is why Sustainable Threads: Collaborative Consumption, Technology, and Green Innovation in Textile is a book that should be of general interest: the textile industry connects us all. Another common dimension that binds humanity together is our pressing need to lead increasingly sustainable processes that permit us to exploit our current productive activities and explore new pathways to innovate while ensuring that future generations will be able to do likewise.

The general aim of this book is to showcase current efforts in sustainability in the textile industry globally and in Spain. Chapter 1 gives an account of the existing research literature on the challenges of sustainability in the fashion industry, offers a review of trends, and identifies gaps for future research. Chapter 2 delves deeper into recent publications on public policy aligned with the Sustainable Development Goals in the principal textile manufacturing regions, namely the European Union, China, and the United States. By leveraging on best practices conducted by fashion and textile firms, the chapter presents case studies that exemplify efforts toward ecologically friendly activities in the major textile economies and outlines key recommendations for firms and governments to follow on in the footsteps of the cases discussed. Chapter 3 focuses on Spanish fashion companies and their efforts to better communicate their achievements in terms of behavioral indicators over the past two decades. Relying on research on one thousand consumers and their perception of the reports and communications issued by 21 Spanish fashion brands, the authors aimed to understand their perception on the credibility of the sales arguments to what extent these brands are associated with sustainable behavior or greenwashing. Chapter 4 presents a case study on Jeanealogia, a Spanish firm that has been innovating in sustainability in jean production processes for over thirty years and which offers their technology and consulting services to top global brands, including Levi Strauss. One of the aims of this chapter is to showcase a business model that can be replicated in different textile manufacturing hubs. Chapter 5 addresses consumer behavior and how fast fashion has had a negative impact on society and the environment. A robust research study, it aims to debunk myths over second-hand clothes ownership and to promote slow fashion and collaborative fashion consumption in Spain, a market which can benefit enormously from these three alternatives to advance on sustainability in the textile and fashion markets. Chapter 6 addresses the role of entrepreneurs in the sustainability in the textile industry and looks into crowdfunding initiative that permit to foresee future trends in this industry. The chapter relies on research that explores ideas from fashion entrepreneurs on mindful consumption, highlights implications, and addresses mindful production as a prerequisite to mindful consumption. Finally, Chap. 7 closes the book by analyzing the circular economy strategies in the luxury fashion industry. More specifically, this chapter presents the case of the renowned firm LVMH and its efforts to promote responsible consumption, based on a qualitatitive and quantitative study that leverages artificial intelligence in content analysis.

The seven chapters comprising this book are authored by renowned researchers in the fields of textile industry and sustainability, and are the product of extensive research efforts. It is our hope that *Sustainable Threads: Collaborative Consumption, Technology, and Green Innovation in Textile* helps inform both consumers and producers globally of current endeavors in sustainability in the textile industry to raise awareness of its importance across all the value chains in the textile industry, and to provide decision makers with a valuable tool for analysis and implementation.

Valencia, Spain
March 2024

José Fernando Gallego-Nicholls
josefernando.gallego@esic.edu

Agustín Carrilero-Castillo
agustin.carrilero@esic.edu

Contents

Chapter 1
Textile Industry and Sustainability: An Integrated Examination

Cristina Blanco González-Tejero, Samuel Ribeiro-Navarrete, Raquel Diaz, and Ruben Furio-Piqueres

1.1 Introduction

Fashion and, in particular, the textile industry have undergone a significant transformation in recent decades. In addition, its link with sustainability, and specifically the circular economy, has been a central topic of analysis in research such as that proposed by Jia et al. (2020).

The textile and fashion industries harm the environment by significantly contributing to water, air, and solid waste pollution, increasing the threats of environmental damage and global warming to living species (Plakantonaki et al., 2023). Research highlights the urgent need to address climate change, particularly in relation to financing and the associated risks, (Shahrour et al., 2023). Thus, this is a dimension that connects all entities integrated into society. In sum, achieving sustainability in our development has been an enduring and significant challenge for humanity (Li et al., 2023).

The evolution of the garment and fashion industry is propelled by a confluence of factors, encompassing technological advances, shifts in consumer preferences, and a heightened environmental consciousness. Consequently, innovation within this

The original version of the chapter has been revised: Three more authors added to the author list.
A correction to this chapter can be found at
https://doi.org/10.1007/978-981-97-5073-3_8

C. Blanco González-Tejero (✉)
Univeristy of Alcala, Madrid, Spain
e-mail: cristina.blancog@uah.es

S. Ribeiro-Navarrete
universidad tecnológica atlantico mediterráneo, UTAMED, Spain

R. Diaz · R. Furio-Piqueres
University of Valencia, Valencia, Spain

J. F. Gallego-Nicholls and A. Carrilero-Castillo (eds.), *Textile Industry and the SDGs*, SDGs and Textiles, https://doi.org/10.1007/978-981-97-5073-3_1

industry is progressively influenced by the dual pillars of sustainability and technological advancements (Larsson, 2018). In that sense, understanding the behaviour of textile companies regarding the use and discharge of chemicals is crucial for designing environmental risk management strategies (Ho & Watanabe, 2020).

The 2030 Agenda has accelerated international cooperation and diplomacy to address global issues. As indicated by Lafont-Torio et al. (2023), the 2030 Agenda includes 169 global goals that provide a global guide for building a fairer world, and, in addition to its ethical and social justice connotations, its achievement increasingly impacts businesses. The influence of the 2030 Agenda on the textile industry is thus particularly relevant due to the environmental and social challenges inherent to this industry. In a constantly growing and developing world, the pursuit of the latest fashion trends and the accelerated consumption cycle has led to an increase in the production and consumption of clothing. However, this unbridled expansion has had significant consequences for the environment and local communities. In this sense, governments have sought to raise awareness and develop policies that make visible the existing needs of society, proposing measures and strategies towards sustainable development (Xie et al., 2021). The fashion industry is of such economic and business relevance that understanding the influence of sustainability in its processes and supply chain has led to a growing interest in academic research. Thus, authors such as Cai and Choi (2020), analyse the existing body of scientific papers in this field. Therefore, this chapter arises with the aim of continuing the exploration of existing research from a new perspective, thus generating new insights and innovative approaches to address the challenges of sustainability in the fashion industry. Thus, a literature review focused on the analysis of the most representative topics related to the textile industry and its sustainability is proposed. Consequently, this chapter is oriented to the review of trends, thus being able to identify gaps to be explored and highlight emerging perspectives that require attention for future research.

This chapter is presented as follows. Firstly, it considers the relevant trends and main developments in the research area, highlighting the research gap. This is followed by an analysis that considers the main scientific documents listed in the Web of Science (WoS) database. After this, the fundamental ideas and key results are detailed. Subsequently, the conclusions and implications of the chapter are presented, detailing new lines of research. Specifically, this literature review allows positioning of the topic of sustainability in the textile industry, considering the main related issues and the relevance of factors and elements involved in manufacturing and consumption processes.

1.2 Theoretical Background

The overconsumption and rapid production of fashion have a significant impact on our planet. From water pollution to overexploitation of natural resources and the generation of non-biodegradable textile waste, the textile industry has become one of the most polluting industries in the world. This has led to an urgent call for the adoption of more sustainable practices. Societies are changing their production and consumption systems to ensure sustainability and the well-being of future generations (Broega et al., 2017).

1.2.1 The Sustainability Challenge

Sustainability is defined in three social, environmental, and economic dimensions, and to achieve full sustainability, processes must consider all three areas (Mahmoudi & Rasti-Barzoki, 2018). Growing consumer interest in durable and functional clothing made in a sustainable manner has created an opportunity for the incorporation of new materials (Yetisen et al., 2016), creating an opportunity for those who design an optimal strategy.

The quest for sustainability has motivated companies to incorporate more environmentally friendly practices into their processes, but the textile industry still faces environmental challenges, such as water pollution and emissions, despite its attempts at recycling (Sharma et al., 2021). Yusuf et al. (2017) emphasise the growing public demand for natural colorants in the textile industry. There are many contaminants in textile wastewater (Li et al., 2019), so the contamination of this industry is of great concern. Thus, Sarkar et al. (2017) underscore the importance of devising environmentally friendly and cost-effective approaches to tackle water pollution stemming from the textile industry. In that sense, the increasing consumer demand for eco-friendly materials has led to a surge in the use of sustainable plant-based products in the textile industry, offering enhanced functionality and environmental benefits (Shahid & Mohammad, 2013b). Thus, companies have incorporated sustainable policies and practices into their supply chains as an essential step to improve operational efficiency, with benefits such as improved environmental, social, and economic performance (Aytekin et al., 2023).

1.2.2 Business

In response to growing global environmental awareness and competition in international markets, textile companies are driven to adopt environmentally responsible practices (Lai et al., 2012). Competitiveness of companies relies on building an efficient team and a collaborative environment, impacting product quality across all industries (Ogunrinde, 2022). Many of them use sustainability as a strategy to gain competitive advantage, differentiate themselves, and improve their public image and reputation (Shiwanthi et al., 2018).

The textile industry, due to its high competitiveness, plays a key role in driving international trade relations in the context of trade integration, as its relative advantage stimulates trade and exchange between countries around the world (Tseng et al., 2022). Thus, it is one of the largest industries and is adopting sustainable management concepts in its supply chain (Lim et al., 2017). Thus, the challenge for the fashion and textile industry has been to proactively address sustainability, driven by pressure from shareholders or consumers seeking corporate social responsibility rather than government regulations (Na & Na, 2015).

Although sustainability is driven by the rapid evolution of Industry 4.0, changing the way businesses operate and optimism about the opportunities it provides for sustainability offers promising opportunities (Khan et al., 2023), the tools available for sustainable fashion design are often complex, and costly. Consequently, adopting green and eco-friendly materials entails an additional cost that increases the investment of organisations in the textile industry (Tumpa et al., 2019). Based on the theoretical framework developed, the following research questions (RQ) are posed:

RQ1: What is the academic relevance of the textile industry in relation to sustainability?

RQ2: Which countries are the main producers of scientific papers in this area?

RQ3: Which terms are associated with the sustainability of the textile industry?

1.3 Methodology and Analysis

In order to address the research questions posed previously, a systematic literature analysis has been developed. This topic has been widely considered by authors such as Harsanto et al. (2023) who already highlighted the demand for green products and systematically analysed some documents linked to the textile industry.

This analysis constitutes a systematic literature review as it adopts a reproducible, scientific, and transparent process (Tranfield et al., 2003). Thus, the search was carried out in the Web of Science database, following the search query: "textile industry" AND ("SDGs" OR "sustainable" OR "green" OR "green economy") selecting the category title or abstract. With these criteria, a total of 874 scientific documents were obtained in August 2023. This is followed by an analysis of the terms broadly considered and the relationship between them. In this case, an analysis of the conceptual structure between the analysed articles is carried out. In particular, the Bibliometrix R package is used as a scientometric software tool. For the visualization of the information, Bibliometrix is used, which allows the visualization of different attributes at the same time in Sankey charts (Linnenluecke et al., 2020).

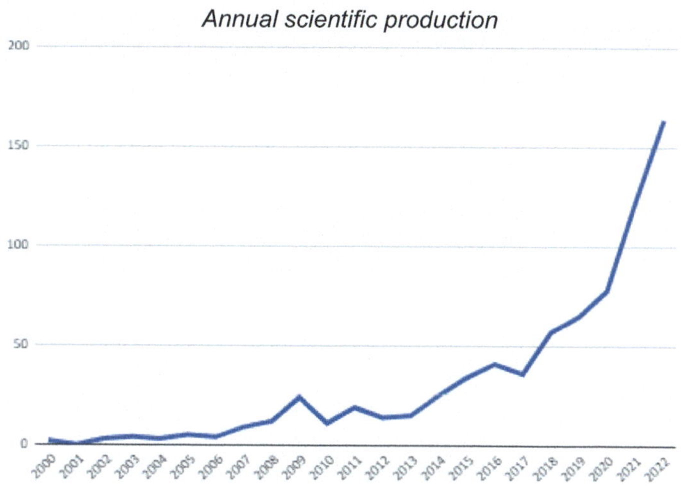

Fig. 1.1 Annual scientific production. *Note* This figure shows the steep increase in scientific output from 2017 on textile industry and sustainability-related terms such as SDGs, green and green economy. *Source* Web of Science, 2023

1.4 Results

1.4.1 Annual Production

Annual scientific production increases notably from 2017, with the highest growth from 2020 onwards. Even so, there was an annual growth of 19.74% in the period analysed. The growth of the world's population is directly responsible for the increase in the production and consumption of textile products (Periyasamy, 2023), which is why issues related to textile sustainability have aroused the interest of many researchers and business leaders. In total, there is an average of 17.72 citations per paper per year and a total of 41,814 references between all documents (Fig. 1.1).

1.4.2 Academic Production by Country

China, a key textile exporter, faces serious energy and environmental challenges due to inefficient use of resources and high levels of pollution (Xu et al., 2018). Thus, it is relevant to consider that the countries that seemingly have the highest risks and present the most challenges in terms of sustainability of textile industrial management are also the ones where there is the highest awareness among researchers. In turn, India, followed by Brazil and Pakistan are the territories with the highest scientific production. Table 1.1 shows that India leads in citations with a total of 2749 citations.

Table 1.1 Citations by country

Country	TC	Average article citations
India	2749	23,30
China	2646	14,70
USA	1104	39,43
Brazil	665	13,85
United Kingdom	607	20,93
México	589	84,14
Turkey	457	11,72
Germany	446	24,78
Tunisia	414	41,40
Poland	383	22,53

Note This table shows a top 10 rank of total citations from the top cited country, India, down to Poland, which is the least cited in this ranking
Source Web of Science, 2023

It is worth noting that territories such as Pakistan, despite having a large scientific volume, do not present the most representative documents in terms of citations. In the top positions in terms of number of citations are the USA and the UK, which are major scientific references (Fig. 1.2).

1.4.3 Terms Analysis

The keywords most frequently used by authors can provide insights into the research focus and areas of interest of scientists and researchers in the field of ecosystem services research (Zhang et al., 2019). In this case, this filter has been used in order to identify the most relevant terms where researchers have put their research focus. In this way, the orange cluster, which refers to the life cycle and circular economy, should be highlighted. Textile companies should be concerned about sustainability in their supply chain due to the need to meet stakeholder objectives (Lim et al., 2017). The purple and green groups refer to the chemical influences on fashion. Textile dyes thus constitute an important sector in the specialty chemicals industry (Kudanga et al., 2011).

In this regard, the maturing public demand in recent times has put pressure on the textile industry for the use of natural dyes, without harmful effects on the environment and aquatic ecosystem, and with more developed functionalities at the same time (Yusuf et al., 2017). The green cluster refers to the production processes of the textile industry. Water consumption and degradation are key concepts to be analysed. Thus, research such as that proposed by Sarkar et al. (2017) which analyses the microbial degradation of water waste from the textile industry should be highlighted. The blue

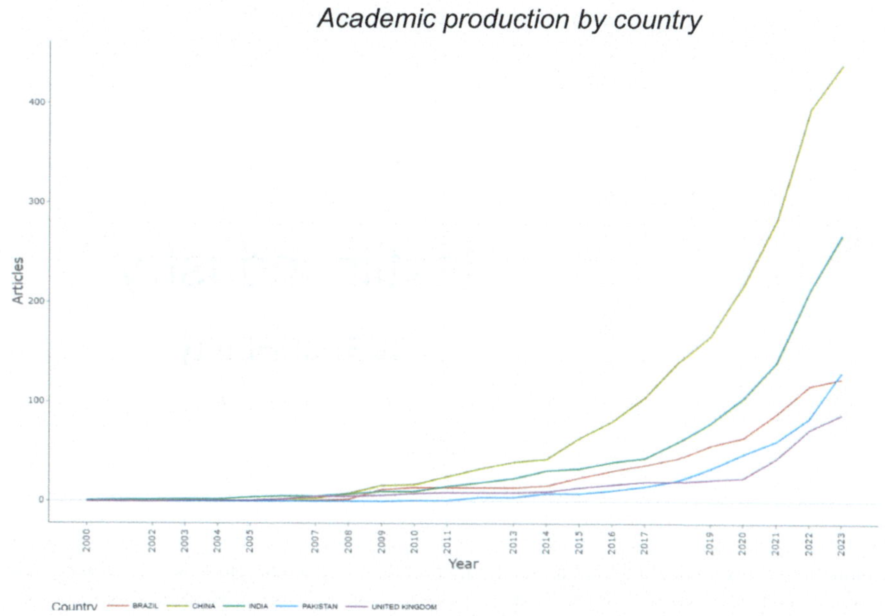

Fig. 1.2 Academic production by country. *Note* This figure shows the increase in academic production on research on the textile industry and sustainability-related terms from 2008 until 2023. Brazil is represented by the orange line, China by the ochre line, India by the light green line, Pakistan by the light blue line, and the United Kingdom by the purple line. *Source* Web of Science, 2023

cluster refers to different textiles, such as cotton and other sustainable materials, as well as highlighting the problem of dyes. Authors such as Shahid and Mohammad (2013a) have highlighted in their research different applications of dyes derived from environmentally friendly plant-based products (Fig. 1.3).

Further analysis of the available information was carried out by analysing the individual words or unigrams available in the summaries of the scientific documents analysed. This analysis made it possible to identify the most relevant and representative words of the content. In this way, these terms help to track current trends in the selected field of analysis. In this case, it is worth emphasising Fig. 1.4, which highlights textile materials, such as "cotton," and production processes. These aspects have been of particular concern to researchers and have become pivotal in their studies, particularly since 2010.

The textile industry plays a key role in the implementation of green chemistry principles (Mishra et al., 2023), thus more and more documents reflect the relevance of the environment and the change of materials and chemicals in textile production processes. Thus, the transition towards circularity in the textile industry, taking into account the large amounts of waste and its immense environmental impact, is challenging but necessary. Therefore, it is evident that terms such as "environmental", "DYE", or "waste" stand out. These terms converge on the need for recycling and environmental awareness of production processes, thus, scholars such as Loo et al.

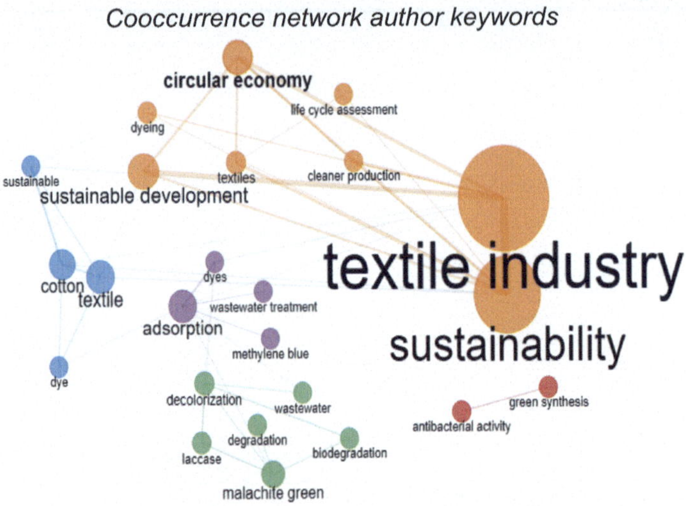

Fig. 1.3 Coocurrence of keywords network. *Note* This figure shows the cooccurrence network of authors and keywords, highlighting the terms prevalent terms adjacent to textile industry and sustainability, namely circular economy, life cycle assessment, cleaner production, dyeing, textiles and sustainable development, in orange. Other terms include different materials, chemical products pertaining to the industry (purple), manufacturing processes (green and blue), and other processes (red). *Source*: author

(2023) point to recycling as a crucial part of the overall circular infrastructure that can facilitate the transition of the textile industry towards a circular and low-carbon economy.

When considering bigrams in the abstracts, a more nuanced understanding can be obtained. In Fig. 1.5, two representative clusters can be observed. One in red, focusing on sustainability and the textile industry, emphasizes terms such as "environmentally friendly" and "circular economy." In the blue cluster, chemical components related to dyes and colour concentration become apparent. Thus, a persistent duality is perceived throughout the entire research, addressing both the chemical components and sustainability in the textile industry. This more detailed approach underscores the importance of comprehending the chemical composition of textile products and how it impacts not only product quality but also its environmental footprint.

1.5 Challenges and Barriers

The textile industry is immersed in constant evolution, a trend reflected in increased research within this area (Desore & Narula, 2018). The internationalization of production processes and the digitalization of business processes mean that companies and entrepreneurs face a multitude of challenges and barriers to the growth

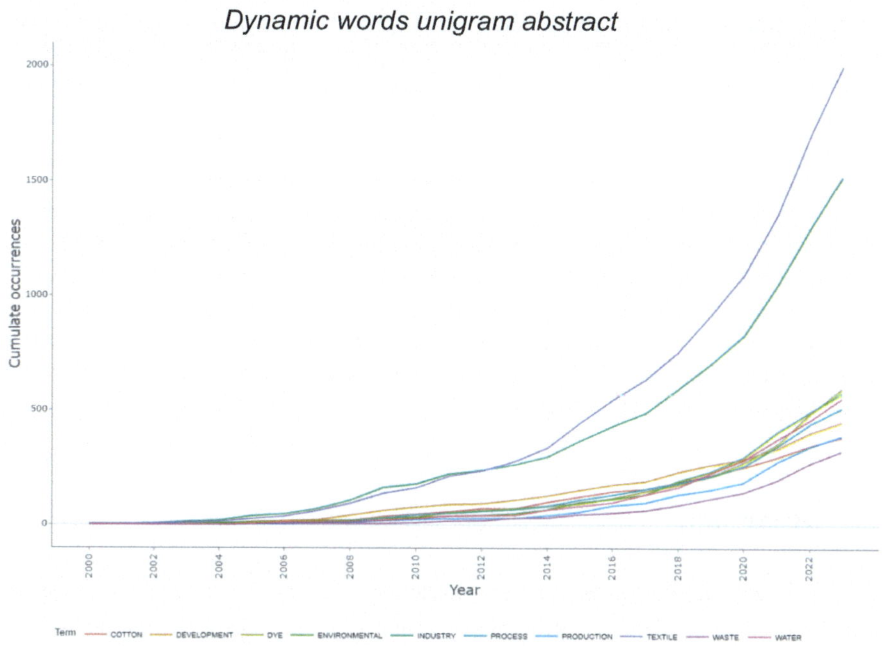

Fig. 1.4 Dynamic words unigram abstract. *Note* This figure shows the increasing interest of researchers in production processes and materials since 2000. *Source* author

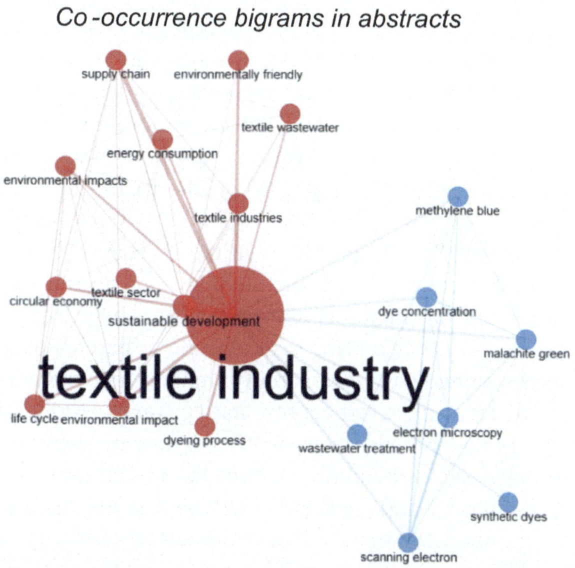

Fig. 1.5 Cooccurrence bigrams in abstracts. *Note* This figure shows two clusters that represent co-occurrence in the textile industry, highlighting the presence of terms such as "environmentally friendly" and "circular economy" (red) and chemical components and processes (blue). *Source* author

and development of their activity. Thus, textile production requires large amounts of resources (water, energy, etc.), so that the overexploitation of resources and polluting elements largely condition the processes developed. On the other hand, textile waste requires correct management, an arduous process in which there are significant obstacles. In addition, business processes have global and complex supply chains, which makes traceability and monitoring of sustainable practices difficult.

This problem is evident in many parts of the world, and the consideration of ethics on the road to more efficient sustainability is significant (Lafont-Torio et al., 2023). On a financial level, it must be considered that adopting sustainable practices in the textile industry requires significant investments in technology and training and can become an obstacle to business management. Finally, environmental and labour regulations are constantly evolving, so textile companies need to be aware of these changing international regulations.

1.6 Discussion

Given the research proposed in this chapter (RQ1), it can be seen that this is a relevant topic that has attracted the attention of different researchers over time, as can be seen in research at different points in time such as Kunamneni et al. (2008) or Mishra et al. (2023). This topic has attracted the attention of researchers such as Sandin and Peters (2018) who have reviewed different publications on the environmental impact of textile reuse and recycling. Accordingly, RQ1 can be addressed indicating that the current relevance is high and continues to grow in view of the academic concern to shed light on the problems of sustainability in the textile industry.

In relation to the second research question (RQ2), it is observed that the regions with the highest scientific production in this field face problems linked to sustainability in their textile industry. In this area, it is worth highlighting research such as that proposed by Xu et al. (2018) who focus their area of action on regions such as China. This observation suggests a crucial intersection between scientific production and real-world problems in these regions. Consequently, the concentration of scientific activity in specific areas gives rise to a recognition of the importance of addressing sustainability concerns within the textile industry. This implies that these areas are not only prolific in generating research, but are also dealing with substantial environmental and sustainability challenges in their textile manufacturing processes.

To select the most representative topics (RQ3), the key terms selected by the authors and the abstracts of the documents were analysed. The first line focuses on business sustainability within the textile industry, highlighting the importance of responsible and environmentally conscious business practices. Key terms such as "environmental impacts," "environmentally friendly," and "textile sector or industry" highlight the relevance of this approach. This alignment between scientific productivity and sustainability issues underscores the importance of research efforts in these regions.

On the other hand, the second line of research is closely linked to chemicals and pollution resulting from textile activities. Keywords such as "dye concentration," "decolorization," and "synthetic dyes" suggest a significant concern for the negative impacts of chemical processes used in the textile industry on the environment. This approach points to the need to address the environmental risks associated with certain chemical components and production practices. Therefore, authors such as Armstrong et al. (2015) make it evident how the garment industry needs innovative business models that can reduce its environmental footprint.

1.7 Conclusions

Through a critical analysis of previous research and a comprehensive review of current trends in sustainability and fashion, this chapter sheds light on the opportunities and challenges facing the industry. In this regard, there has been a growing interest since 2020. The most representative topics refer to sustainability in the textile industry's supply chain, followed by the use of new natural dyes, which would allow the industry's production processes to be adjusted and the waste derived from this manufacturing to be considered. Essentially, exploring the circular economy and the life cycle of textile products has emerged as pivotal areas of research. It is crucial to heighten awareness regarding the environmental impact and sustainability of the textile industry.

The link between the highlighted topics and thematic lines reveals an inherent complexity in the research, where business sustainability and the management of production processes that require chemicals are intertwined. Consequently, the adoption of sustainable practices in the textile industry involves not only business considerations, but also the proper management of chemicals to minimize adverse impacts on the environment. Thus, it can be concluded that the textile industry faces significant challenges and that research on these issues is essential to drive more responsible practices and promote the adoption of sustainable technologies and approaches throughout the textile supply chain.

1.8 Implications

This chapter is useful for researchers and business leaders as it identifies prevailing trends and approaches which, on a theoretical level, can help to better understand the current direction of the textile industry and the key elements where the focus should be placed in order to contribute to the sustainability of the textile industry. Thus, as awareness of the environmental impact of the industry increases, opportunities arise to innovate and find solutions that allow a balance between fashion, economy, and the protection of the natural environment.

At a theoretical level, this research provides a summarised overview of the main contributions in the field of analysis, serving as a starting point and synthesis for future research. Considering the practical implications, this chapter enables leaders in regions with significant environmental impact and concern to formulate sustainability strategies based on the results obtained from scientific articles and documents. Consequently, informed decisions for the future can be made based on relevant research in the field. Additionally, by examining the impact of the textile industry, it will be possible to foster international relationships and collaborations, thereby supporting joint action plans.

1.9 Limitations

This chapter is not without limitations. Thus, due to the growing trend in the number of research projects in this field, the contents need to be updated in order to analyse whether the research gaps are being filled and whether new opportunities arise to continue exploring. In addition, an analysis of the existing literature has been carried out from an analytical point of view of the topics considered, thus being able to consider new bibliometric metrics that allow us to analyse the most relevant academics. In this case, the Web of Science database has been used, and it may be relevant to extend the considerations with databases such as Scopus. Consequently, this chapter allows us to continue with the analysis and reflect on the current situation of sustainable practices in the textile industry.

1.10 Research Agenda

This topic has garnered the attention of numerous academics, underscoring the importance of continued exploration. The objective is to delve into the best management practices for the textile industry and identify pathways towards a more sustainable industry. In this context, it is essential to raise awareness in society about these aspects. Along these lines, these questions can help guide future research and explore new dimensions:

- How can technologies influence the direction of research in this field?
- What is the impact of sustainability and corporate social responsibility initiatives on design and production practices in the fashion industry?
- How can small and medium-sized fashion enterprises (SMEs) effectively address sustainability challenges in their operations and supply chains?
- What is the role of consumer education and awareness in promoting sustainable fashion?

References

Armstrong, C. M., Niinimäki, K., Kujala, S., Karell, E., & Lang, C. (2015). Sustainable product-service systems for clothing: Exploring consumer perceptions of consumption alternatives in Finland. *Journal of Cleaner Production, 97*, 30–39.

Aytekin, A., Okoth, B. O., Korucuk, S., Karamaşa, Ç., & Tirkolaee, E. B. (2023). A neutrosophic approach to evaluate the factors affecting performance and theory of sustainable supply chain management: Application to textile industry. *Management Decision, 61*(2), 506–529.

Broega, A. C., Jordão, C., & Martins, S. B. (2017). Textile sustainability: Reuse of clean waste from the textile and apparel industry. In *IOP Conference Series: Materials Science and Engineering* (Vol. 254, No. 19, p. 192006). IOP Publishing. (2017, October).

Cai, Y. J., & Choi, T. M. (2020). A united nations' sustainable development goals perspective for sustainable textile and apparel supply chain management. *Transportation Research Part E: Logistics and Transportation Review, 141*, 102010.

Desore, A., & Narula, S. A. (2018). An overview on corporate response towards sustainability issues in textile industry. *Environment, Development and Sustainability, 20*, 1439–1459.

Harsanto, B., Primiana, I., Sarasi, V., & Satyakti, Y. (2023). Sustainability innovation in the textile industry: A systematic review. *Sustainability, 15*(2), 1549.

Ho, H. T. T., & Watanabe, T. (2020). The dynamics of textile firm responses to the restriction of endocrine disrupting surfactants: A Vietnam case study. *Journal of Risk Research, 23*(4), 504–521.

Jia, F., Yin, S., Chen, L., & Chen, X. (2020). The circular economy in the textile and apparel industry: A systematic literature review. *Journal of Cleaner Production, 259*, 120728.

Khan, S. A. R., Tabish, M., & Zhang, Y. (2023). Embracement of industry 4.0 and sustainable supply chain practices under the shadow of practice-based view theory: Ensuring environmental sustainability in corporate sector. *Journal of Cleaner Production, 398*, 136609.

Kudanga, T., Nyanhongo, G. S., Guebitz, G. M., & Burton, S. (2011). Potential applications of laccase-mediated coupling and grafting reactions: A review. *Enzyme and Microbial Technology, 48*(3), 195–208.

Kunamneni, A., Ghazi, I., Camarero, S., Ballesteros, A., Plou, F. J., & Alcalde, M. (2008). Decolorization of synthetic dyes by laccase immobilized on epoxy-activated carriers. *Process Biochemistry, 43*(2), 169–178.

Lafont-Torio, J., Martín, J. M. M., Fernández, J. A. S., & Ribeiro-Soriano, D. (2023). Perceptions of progress toward achieving the sustainable development goals: Insights from cooperative managers. *Sustainable Technology and Entrepreneurship*, 100055.

Lai, R. S., Hsu, L. L., & Chen, J. C. (2012). Green supply chain management systems: A case study in the textile industry. *Human Systems Management, 31*(2), 111–121.

Larsson, J. K. J. (2018). Digital innovation for sustainable apparel systems: Experiences based on projects in textile value chain development. *Research Journal of Textile and Apparel, 22*(4), 370–389.

Li, Q., Liao, Z., Fang, X., Wang, D., Xie, J., Sun, X., Wang, L., & Li, J. (2019). Tannic acid-polyethyleneimine crosslinked loose nanofiltration membrane for dye/salt mixture separation. *Journal of Membrane Science, 584*, 324–332.

Li, X., Wu, T., Zhang, H. J., & Yang, D. Y. (2023). National innovation systems and the achievement of sustainable development goals: Effect of knowledge-based dynamic capability. *Journal of Innovation & Knowledge, 8*(1), 100310.

Lim, M. K., Tseng, M. L., Tan, K. H., & Bui, T. D. (2017). Knowledge management in sustainable supply chain management: Improving performance through an interpretive structural modelling approach. *Journal of Cleaner Production, 162*, 806–816.

Linnenluecke, M. K., Marrone, M., & Singh, A. K. (2020). Conducting systematic literature reviews and bibliometric analyses. *Australian Journal of Management, 45*(2), 175–194.

Loo, S. L., Yu, E., & Hu, X. (2023). Tackling critical challenges in textile circularity: A review on strategies for recycling cellulose and polyester from blended fabrics. *Journal of Environmental Chemical Engineering,* 110482.

Mahmoudi, R., & Rasti-Barzoki, M. (2018). Sustainable supply chains under government intervention with a real-world case study: An evolutionary game theoretic approach. *Computers & Industrial Engineering, 116*, 130–143.

Mishra, K., Siwal, S. S., Nayaka, S. C., Guan, Z., & Thakur, V. K. (2023). Waste-to-chemicals: Green solutions for bioeconomy markets. *Science of the Total Environment*, 164006.

Na, Y., & Na, D. K. (2015). Investigating the sustainability of the Korean textile and fashion industry. *International Journal of Clothing Science and Technology, 27*(1), 23–33.

Ogunrinde, A. (2022). The effectiveness of soft skills in generating dynamic capabilities in ICT companies. *ESIC Market, 53*(3), e286–e286.

Periyasamy, A. P. (2023). Microfiber emissions from functionalized textiles: Potential threat for human health and environmental risks. *Toxics, 11*(5), 406.

Plakantonaki, S., Kiskira, K., Zacharopoulos, N., Chronis, I., Coelho, F., Togiani, A., Kalkanis, K., & Priniotakis, G. (2023). A review of sustainability standards and ecolabeling in the textile industry. *Sustainability, 15*(15), 11589.

Sandin, G., & Peters, G. M. (2018). Environmental impact of textile reuse and recycling–A review. *Journal of Cleaner Production, 184*, 353–365.

Sarkar, S., Banerjee, A., Halder, U., Biswas, R., & Bandopadhyay, R. (2017). Degradation of synthetic azo dyes of textile industry: A sustainable approach using microbial enzymes. *Water Conservation Science and Engineering, 2*, 121–131.

Shahid, M., & Mohammad, F. (2013a). Green chemistry approaches to develop antimicrobial textiles based on sustainable biopolymers a review. *Industrial & Engineering Chemistry Research, 52*(15), 5245–5260.

Shahid, M., & Mohammad, F. (2013b). Perspectives for natural product based agents derived from industrial plants in textile applications–A review. *Journal of Cleaner Production, 57*, 2–18.

Shahrour, M. H., Arouri, M., & Lemand, R. (2023). On the foundations of firm climate risk exposure. *Review of Accounting and Finance.* https://doi.org/10.1108/RAF-05-2023-0163

Sharma, S., Prakash, G., Kumar, A., Mussada, E. K., Antony, J., & Luthra, S. (2021). Analysing the relationship of adaption of green culture, innovation, green performance for achieving sustainability: Mediating role of employee commitment. *Journal of Cleaner Production, 303*, 127039.

Shiwanthi, S., Lokupitiya, E., & Peiris, S. (2018). Evaluation of the environmental and economic performances of three selected textile factories in Biyagama Export Processing Zone Sri Lanka. *Environmental Development, 27*, 70–82.

Tranfield, D., Denyer, D., & Smart, P. (2003). Towards a methodology for developing evidence-informed management knowledge by means of systematic review. *British Journal of Management, 14*, 207–222.

Tseng, M. L., Bui, T. D., Lim, M. K., Fujii, M., & Mishra, U. (2022). Assessing data-driven sustainable supply chain management indicators for the textile industry under industrial disruption and ambidexterity. *International Journal of Production Economics, 245*, 108401.

Tumpa, T. J., Ali, S. M., Rahman, M. H., Paul, S. K., Chowdhury, P., & Khan, S. A. R. (2019). Barriers to green supply chain management: An emerging economy context. *Journal of Cleaner Production, 236*, 117617.

Xie, H., Wen, J., & Choi, Y. (2021). How the SDGs are implemented in China—A comparative study based on the perspective of policy instruments. *Journal of Cleaner Production, 291*, 125937.

Xu, C. K., Cheng, H., & Liao, Z. J. (2018). Towards sustainable growth in the textile industry: A case study of environmental policy in China. *Polish Journal of Environmental Studies, 27*(5).

Yetisen, A. K., Qu, H., Manbachi, A., Butt, H., Dokmeci, M. R., Hinestroza, J. P., Skorobogatiy, M., Khademhossein, A., & Yun, S. H. (2016). Nanotechnology in textiles. *ACS Nano, 10*(3), 3042–3068.

Yusuf, M., Shabbir, M., & Mohammad, F. (2017). Natural colorants: Historical, processing and sustainable prospects. *Natural Products and Bioprospecting, 7*, 123–145.

Zhang, X., Estoque, R. C., Xie, H., Murayama, Y., & Ranagalage, M. (2019). Bibliometric analysis of highly cited articles on ecosystem services. *PLoS ONE, 14*(2), e0210707.

Chapter 2
Exploratory Analysis of Best Practices in Green Innovation and the SDGs in Major Textile Economies

Abel Monfort, Andrés Gómez Funes, and José Manuel Mas-Iglesias

2.1 Textile Industry and SDGs

The textile industry is essential in the global economy, contributing to trade, employment, investment, and income worldwide (Arruda-Filho et al., 2019). However, the industry faces sustainability challenges, particularly environmental impact and pollutant emissions. Achieving green and sustainable economic development in the textile industry requires a balanced approach to economic growth and pollution control (Qingfen et al., 2023). The adverse effects of the textile industry on sustainability have drawn attention to the need to integrate the United Nations Sustainable Development Goals (SDGs) into business communication, emphasizing the industry's potential for both development and sustainability (Olofsson & Mark-Herbert, 2020).

Conceptually, sustainable fashion refers to companies' commitment to their products' origin, manufacturing process, logistics, or marketing, emphasizing respect for the environment and principles related to the circular economy. Consequently, investment in the sustainable development of the textile industry is crucial to achieving the SDGs, particularly those most closely related to environmental issues. This includes adopting advanced technologies, transforming business models, and the implementation of economic initiatives and policies supporting sustainable development (Liu et al., 2021). In addition, the industry is transitioning to sustainable fashion by using environmentally friendly materials such as organic cotton and recycled polyester fiber (Fadara & Wong, 2022). Business leadership also plays a vital role in promoting

---

A. Monfort (✉) · A. Gómez Funes · J. M. Mas-Iglesias
ESIC University, ESIC Business & Marketing School, Madrid, Spain
e-mail: abel.monfort@esic.university

A. Gómez Funes
e-mail: andres.gomez@esic.university

J. M. Mas-Iglesias
e-mail: josemanuel.mas@esic.university

J. F. Gallego-Nicholls and A. Carrilero-Castillo (eds.), *Textile Industry and the SDGs*, SDGs and Textiles, https://doi.org/10.1007/978-981-97-5073-3_2

sustainable development within the textile industry, particularly in achieving SDG 8 (decent work and economic growth) (Shehri et al., 2022).

In this context, fashion consumers face a severe dilemma, as the ideal of sustainability seems inconsistent with the essence of modern fashion: frugal, trendy, ever-changing, and extreme. On the one hand, fashion companies focus on designing and producing new collections between two and eight times a year. When brands can no longer sell products at the total price, they apply discounting strategies. Garments can go to an outlet store or be sold through direct sales, promotions, and online discount platforms, among other ways (Díaz-Soloaga, 2019). At this point, brands aim to eliminate all the accumulated stock at any price. This process occurs quickly and is promoted with constant messages to consumers. Brands develop an incessant communication activity to get their products to reach consumers several times during the same year (Chan & Wong, 2012). Surprisingly, some of these same brands are concerned about sustainable processes and, in many cases, are committed to the circular economy. In their commercial communications, they discuss reducing their ecological footprint and tackling the pollution caused by their production process. This seemingly irreconcilable logic deserves the attention of academics and experts. Furthermore, it is worth looking at what the countries in which these brands have been created are doing to promote sustainable growth.

From a broader point of view, it is known that business management of sustainability acquires particular relevance and requires multiple institutional logics, considering stakeholders and integrating multiple dimensions to deliver value (Schneider & Claußß, 2020). Managers must take responsibility for the sustainability policy, and incentives linked to its measurement must be established (Eccles et al., 2014). For that reason, business schools are increasingly promoting education based on ethics, sustainability, and social responsibility (Fornes et al., 2019), new forms of financing are being sought to meet the challenges of the 2030 Agenda (Méndez-Suárez et al., 2020) and market players are putting pressure on the most polluting companies (Boiral et al., 2012; Okereke & Russel, 2010). Sustainable innovation has garnered significant attention in recent years, with scholars and practitioners recognizing the importance of effectively managing and communicating such innovation. Monfort and Mas-Iglesias (2021) noted that this trend is set to continue, underscoring the need for continued research and analysis in this area.

According to the literature, the sector's transformation toward sustainability encompasses different approaches. Different examples arise, such as the proposal of an evaluation index system to help textile and clothing companies understand the transition phase of the industry and achieve sustainable development (Niu & Zhu, 2021), the integration of environmental protection concepts in textile education to promote green practices in the sector (Pu et al., 2021); promoting green textiles and encouraging the use of sustainable materials to reduce adverse effects on the environment (Sadeka & Hammash, 2023) or restructuring human resource strategies and industry curricula to train socially responsible and eco-efficient human resources in line to promote sustainable lifestyles worldwide (Jain, 2022). In addition, incremental sustainable organizational change is being explored within the textile and apparel industry to promote sustainability in strategic management practices (Patora-Wysocka & Sułkowski, 2019).

The industry's transition to sustainability involves a multifaceted approach, encompassing technological advances, policy support, educational reforms, and collaborative research efforts to drive sustainable development and mitigate environmental impact.

SDGs and green sustainability in textile economies.

The textile industry actively engages in sustainable practices aligned with the Sustainable Development Goals (SDGs), focusing on green supply chain management, eco-friendly textiles, and stakeholder engagement, thus applying environmental sustainability measures and eco-design to promote sustainable practices (Islam et al., 2022). However, challenges remain as the textile industry grapples with balancing economic development and pollutant emissions, highlighting the importance of addressing these issues to achieve sustainable economic growth (Qingfen et al., 2023). In addition, the industry is exploring eco-friendly textiles as sustainable alternatives to conventional materials, aiming to mitigate environmental impact (Sadeka & Hammash, 2023).

In this context, major exporters are making great efforts for green conversion. In China, they emphasize the importance of green development and sustainable practices in their textile industry, recognizing the industry's role in promoting related sectors and stimulating domestic demand (Li & Yi, 2019). Sri Lanka and Bangladesh are advancing sustainable practices within their garment industries, focusing on responsible fashion, waste management, and proactive business expansion to align with green principles (Shamsuzzaman et al., 2023). These are just a few examples of what is being done internationally.

Despite these efforts, challenges such as microfiber pollution and textile waste generation persist, requiring comprehensive approaches to address sustainability in the textile and fashion industry (Liu et al., 2021). However, advances in green chemistry and bio-based synthetic textile fibers offer promising avenues for sustainable textile production and waste regeneration (Sun et al., 2021; Tian et al., 2022). Furthermore, the significant contribution of the textile industry to water pollution underscores the urgency of addressing sustainability issues (Saini, 2023).

The objective of this chapter is to scrutinize the worldwide endeavors being undertaken in connection with the sustainable development goals (SDGs) that are interrelated to environmental sustainability (Boström, & Micheletti, 2016). The ensuing case studies have honed in on territories or nations with a noteworthy degree of textile exports across the globe and which have acknowledged fashion brands to investigate how the endeavors of these nations toward a more ecologically friendly textile economy are mirrored in the fashion industry.

2.1.1 European Union

The European Union (E.U.) has been actively investing in eco-innovations related to the Sustainable Development Goals (SDGs) in the textile industry (Yurdakul & Kazan, 2020). One of the initiatives is the European Clothing Action Plan (ECAP), aimed at significantly improving the sustainability of textiles throughout their life cycle, from design to end of use (Jacometti, 2019; Moorhouse & Moorhouse, 2017). This initiative aligns with E.U.'s efforts to implement a circular economy model focusing on waste reduction, addressing the global concern of textile waste (Jacometti, 2019). In addition, the E.U. has played a crucial role in promoting eco-innovations in companies, conditioning companies' accessibility to national and E.U. funding on investment in these innovations (Siedschlag et al., 2019).

These initiatives have aided and driven transformations in representative EU fashion companies. A couple of examples are highlighted below, such as H&M and Inditex. Swedish firm H&M has invested in adopting renewable energy sources to power its operations. H&M has implemented solar and wind power technologies in its facilities, covering retail stores and distribution centers. This initiative has enabled H&M to reduce its carbon footprint, aligning with SDG 13 (Climate Action), and reflects its commitment to decarbonizing the textile supply chain. In 2022, around 92% of H&M's total electricity consumption came from renewable sources. This decreased from the previous year when the company's renewable energy use stood at 95%. The company's goal is to reach 100% renewable energy by 2030 (Fig. 2.1).

The organization has undertaken several measures to address its carbon footprint and reduce value chain initiatives. These measures include strengthening the fashion supply chain to leverage clean energy, energy efficiency, and sustainable materials as part of the Green Fashion Initiative. Furthermore, the organization has recently emerged as the lead funder of the $250 million Fashion Climate Fund, which aims to promote sustainability in the fashion industry. In addition, the organization has

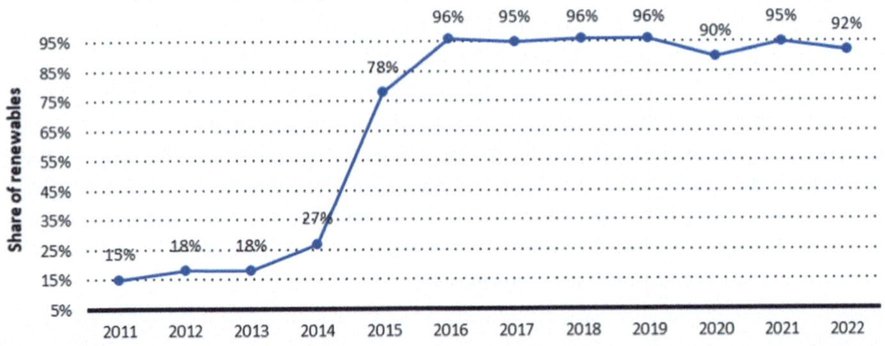

Fig. 2.1 Share of renewables in total electricity consumption by H&M from 2011 to 2022 *Source* H&M (2022)

launched a Sustainable Supplier Facility in collaboration with the consultancy Guidehouse, which provides suppliers with financing related to decarbonization. Moreover, the organization has implemented an internal carbon price to drive behavioral changes in purchasing, design, marketing, production, and logistics. The company, according to Leyla Ertur, Head of Sustainability H&M Group, said: "We are dedicated to contributing to the limitation of the global temperature increase to 1.5 °C and keeping our climate strategy aligned with the latest science-based target setting frameworks is crucial for us to create impact beyond our own organization and value chain" (Segal, 2022).

Another example is Inditex, the Spanish-based multinational fashion conglomerate founded by Amancio Ortega. The company owns popular brands such as Zara, Pull & Bear, Massimo Dutti, and Bershka. Inditex operates in the fast fashion sector and is known for its efficient business model and ability to adapt to market trends quickly. It has historically been one of the most profitable companies in the fashion industry, generating significant profits over the years.

In 2001, the company became a signatory to the United Nations Global Compact and, since then, has sought to demonstrate its commitment to the environment. Its objectives include the following: The year 2023 ended it seeking 100% elimination of single-use plastics for its customers and 100% use of organic, recycled, or Better Cotton. To this end, they are reusing and recycling all hangers and labels by the end of 2023, almost eliminating single-use plastic for their customers. Thus, they replace plastic bags in their stores with recycled paper bags and encourage reusable bags by charging for recycled paper bags. The equivalent value of the proceeds is invested entirely in environmental initiatives.

In addition, Inditex helps its customers extend the life of their garments and move toward a circular economy model. With this, the company launched Zara Pre-Owned in 2022, a pioneering platform that offers repair services, private-to-private sales, and donation of garments from its Zara brand. It also allows its customers to deposit their used garments, footwear, and accessories in the Group's stores in collaboration with organizations that work on their reuse and recycling.

Zara Pre-owned is a platform offering repair services, customer-to-customer sales, and donated used garments. This service is currently available in the United Kingdom and France. The company is committed to making it operational in all strategic markets by 2025. Among its services, it has a network of workshops that work to find the best solution for damaged garments. In addition, it has a service for reselling Zara garments or buying second-hand items to make them last longer and extend their useful life. Finally, within Zara Pre-owned, a program has been launched to collect used clothing, footwear, and accessories. The project aims to promote the recovery of used products to extend their useful life while positively impacting the community. All items collected are delivered to local organizations, which sort them according to their condition and quality. These products are donated to people at risk of social exclusion, sold in second-hand stores, or recycled. The proceeds are used to finance the social projects of these organizations, such as the Red Cross, Le Relais, Caritas, and the China Environmental Protection Foundation, among others.

2.1.2 China

China has been undergoing a significant economic transformation from an export- and investment-driven model to a consumption- and innovation-driven one. This structural change and industrial upgrading have created new opportunities to achieve sustainable growth (Xu & Wang, 2019). Textile industry innovation in China has become a crucial issue at the core of sustainable industrial development (Zhang et al., 2020). The country faces pressure to reduce energy consumption, and the industry has been the focus of water conservation and waste reduction in China's Thirteenth Five-Year Plan (2016–2020) (Wang et al., 2017), as well as in the current Plan.

The Chinese government has formulated policies to guide enterprises in introducing advanced environmental protection, energy, and operating technologies, emphasizing the need for cleaner production. In addition, numerous technologies are emerging to reduce water use and pollution in the Chinese textile industry, including those promoted by the China National Textile and Apparel Council (Chen et al., 2019). However, some studies have analyzed the interaction between environmental efficiency and production efficiency in the textile industry in China, highlighting the need to balance these aspects for sustainable development (Jiang et al., 2016). The low-carbon development mode has been explored in the context of China's textile and garment export business, emphasizing implementing a low-carbon economic environment in this sector (Chang & Zhu, 2022). In addition, the Chinese textile industry faces pollution-related challenges in the production and sales processes, making it necessary to focus on environmental protection and low-carbon measures (Zhu & Hu, 2016). In this context, some examples of the country's leading apparel marketing companies are shown below.

One leading company in the Chinese textile sector that has shown a prominent interest in sustainability is Anta Sports. Anta Sports Products Limited is a Chinese sports apparel and footwear company. Anta is one of the largest sportswear companies in China and has experienced significant growth in recent years. The company was founded in 1994 in Fujian, China, and has expanded its operations domestically and internationally. Anta specializes in manufacturing and selling sports footwear, apparel, and accessories. In addition to the leading brand "Anta," the company owns other well-known brands such as FILA, Descente, and Sprandi.

Anta Group and the World Wide Fund for Nature (WWF) announced in 2023 that Anta Group would become WWF's first international partner in China's textile and sportswear industry. This agreement is an extension of the one they developed in March 2020 to promote "Earth Hour." In the collaboration, they are going to carry out the following activities: improve industrial water and energy use, industrial waste emissions, and supply chain management of at least 3,000 textile suppliers; reduce energy consumption per unit of industrial value added by 30% and water consumption per unit of industrial value added by 22.5%. In addition, they will reduce total emissions of major water pollutants cumulatively by 15% to reduce the impact of the textile industry on the environment and reduce freshwater, climate, and its footprint on the ecosystem.

Another relevant example in China is JD.com, also known as Jingdong. It is one of the largest e-commerce and online retail companies in China. It was founded by Liu Qiangdong in 1998 in Beijing, China. Initially, the company focused on selling electronics but eventually expanded to include various product categories, including fashion. Hence, it is worth highlighting in this work. Unlike other e-commerce giants such as Alibaba, which is based on a marketplace model (such as Alibaba and Taobao), JD.com operates a direct selling business model. This means it buys and stocks inventory directly and manages its logistics and delivery.

JD.com has taken several steps to integrate sustainable practices into its operations and contribute to environmental and social responsibility. For example, its commitment to sustainable packaging to reduce waste stands out. The company has implemented smaller, lighter packaging and recyclable materials to minimize environmental impact. JD.com has adopted three-layer cartons (which are typically five layers) for 95% of its cartons, while the average rate in China is 60–70%. Similarly, JD.com has succeeded in decreasing the thickness of plastic delivery bags from 60 to 70 μm to 40 to 50 μm. In 2015, JD.com proposed the "thin tape" concept to the industry to replace 60-mm packaging tapes with 45-mm tapes, based on the company's repeated and verified tests in its logistics business. Now, JD.com's "thin tape" has been included in China's standards for the postal industry and is used by third-party logistics companies.

2.1.3 United States

The U.S. textile industry increasingly focuses on sustainability, particularly on sustainable supply chain management, sustainable product strategies, and sustainable materials. Consequently, it can be argued that the textile industry's sustainability efforts are influenced by global initiatives such as the Sustainable Development Goals, which drive the industry to focus on environmental impact and sustainable practices (Wood & Redfern, 2022). This shift is driven by the recognition of the environmental impact of textile production and the need for sustainable development (Fadara & Wong, 2022; Liu et al., 2021; Oelze, 2017; Shen et al., 2017). The industry also faces challenges in implementing circular economy practices and addressing microfiber pollution, which requires investing in advanced technologies and transforming business models (Alves et al., 2021; Chen et al., 2021). In addition, adopting sustainable manufacturing practices and integrating sustainability into human resource development has become crucial for the sustainable transformation of the industry (Jain, 2022; Patora-Wysocka & Sułkowski, 2019). Adopting green technologies, such as blockchain, for waste management and applying deep-learning-based risk analysis methods for foreign investments are increasingly crucial for the industry (Dursun et al., 2022). In addition, the integration of sustainability into curricula and the development of traceability systems are contributing to the sustainable transformation of the industry (Kumar et al., 2017; Wood & Redfern, 2022).

The industry recognizes the importance of assessing and managing the social responsibility risks associated with overseas investments (Cao & Yao, 2021). First, Levi Strauss & Co. stands out. The iconic American company was founded in 1853 and is known worldwide for its influence on denim fashion. It is a pioneer in manufacturing jeans and casual wear, standing out for its Levi's brand, which has endured through generations.

Levi's has invested in energy efficiency and adopted renewable energy sources to reduce its environmental impact. The company has implemented measures to improve sustainability in producing its products and has set specific targets to reduce greenhouse gas emissions. Levi Strauss & Co. has made several efforts in the area of sustainability. Here are some highlights. Among others: the efficient use of water. To this end, it has implemented programs to reduce water use in its manufacturing processes. They have adopted technologies and practices that allow water conservation in the production of their denim garments, one of the areas with the most significant environmental impact in the textile industry. The company has also launched products made from recycled materials, including denim garments from recycled plastic bottles.

2.2 Conclusions and Practical Implications

This analysis delves into the best practices in green innovation and their connection to the Sustainable Development Goals (SDGs) in major textile economies. It highlights the importance of addressing sustainability challenges within the clothing industry, significantly contributing to global trade, employment, and income (McNeill & Moore, 2015). However, harmful environmental impacts such as pollution and emissions must also be addressed. Thus, the link between fashion and the SDGs is vital (Patyal et al., 2022; Vijeyarasa & Lui, 2021).

The global focus is on transitioning to sustainable fashion, with textile companies seeking to adopt greener practices. This shift involves investments in advanced technologies, transformation of business models, and economic and political initiatives to support sustainable development (Negrete & López, 2020). Corporate sustainability management is crucial, with managers being held accountable for sustainable policies, and market pressure to address environmental issues is increasing (Pérez et al., 2021).

To illustrate how major textile economies respond to these challenges, this analysis provides specific examples from the European Union, China, and the United States. The E.U. has implemented initiatives such as the European Clothing Action Plan to improve sustainability in the textile supply chain. In China, leading companies such as Anta Sports and JD.com are collaborating with environmental organizations to reduce their environmental impact. In the United States, iconic companies such as Levi Strauss & Co. are investing in energy efficiency and adopting more sustainable practices in producing their products. This highlights the importance of assessing and managing the social responsibility risks associated with overseas investments.

In conclusion, this analysis stresses the urgent need to address environmental challenges in the textile industry and the significant efforts being made worldwide to achieve a transition to sustainability. Adopting greener practices is both an ethical imperative and an innovative business strategy, especially in a world where consumers are increasingly aware of sustainability.

Implications for Policy Makers:

To effectively address the textile industry's environmental and social challenges, policymakers must prioritize the implementation of regulations and standards that promote sustainable practices. This includes incentivizing eco-friendly materials, encouraging responsible waste management, and supporting initiatives promoting fair labor practices. Policymakers can drive positive change and create a sustainable future for the textile industry by actively engaging with stakeholders and fostering collaboration between industry players.

For example, policymakers could introduce a regulation requiring textile manufacturers to use a certain percentage of recycled or organic materials. This would incentivize companies to source sustainable materials and reduce their reliance on virgin resources. Additionally, policymakers could implement a waste management system that encourages recycling and proper disposal of textile waste, reducing the industry's environmental footprint and promoting circularity.

2.3 Key Recommendations

1. **Incentivizing Sustainable Practices**: Encouraging investment in green technologies and sustainable business models through subsidies, tax incentives, or grants. Implementing strict regulations and standards: Establishing mandatory guidelines for using recycled or organic materials in textile production and enforcing penalties for non-compliance. Creating awareness and education campaigns: Educating consumers about the textile industry's environmental impact and promoting the benefits of buying sustainable and recycled products. Collaborating with stakeholders: Working closely with fashion brands, manufacturers, and consumers to develop innovative solutions and strategies for a more sustainable and circular fashion industry.
2. **Regulatory Frameworks**: Establishing stringent environmental regulations and standards to ensure compliance with sustainable practices. This includes setting limits on carbon emissions, water usage, and waste disposal. Additionally, implementing strict rules on the use of hazardous chemicals and promoting the adoption of eco-friendly alternatives in the production process is crucial. Lastly, monitoring and enforcing penalties for non-compliance is essential to ensure that all stakeholders adhere to sustainable practices and contribute to a more environmentally friendly fashion industry. By creating awareness, collaborating with stakeholders, and implementing regulatory frameworks, we can pave the way for a

more sustainable and circular fashion industry that benefits both the environment and consumers.

3. **Public–Private Partnerships**: Fostering collaborations between government bodies and private organizations to innovate in sustainable textile manufacturing and waste management is another crucial step in creating a more environmentally friendly fashion industry. These partnerships can develop and implement innovative solutions that address the industry's environmental impacts by combining resources, expertise, and knowledge. This can include initiatives such as investing in research and development for sustainable materials, promoting efficient waste management practices, and supporting the adoption of clean and energy-efficient technologies. Through these collaborations, the fashion industry can make significant progress toward achieving sustainability goals and reducing its environmental footprint.
4. **Consumer Awareness Campaigns**: Supporting initiatives that educate consumers about the benefits of sustainable fashion, thereby driving market demand for eco-friendly products. Consumer awareness campaigns drive market demand for eco-friendly products in the fashion industry. By educating consumers about the benefits of sustainable fashion, such as reduced carbon emissions and water usage, these initiatives can create a shift in consumer behavior toward more environmentally conscious choices. Additionally, by increasing demand for sustainable products, these campaigns can incentivize fashion brands to prioritize sustainability in their production processes, reducing the industry's overall environmental footprint.
5. **Global Cooperation**: Advocating for international agreements and collaborations to address the transboundary nature of environmental challenges in the textile industry. Global cooperation is essential in addressing the textile industry's transboundary nature of environmental challenges. By advocating for international agreements and collaborations, countries can work together to set global standards and regulations for sustainable fashion practices. This can help ensure that signatory countries are held accountable for their environmental impact and encourage a more consistent and widespread adoption of sustainable practices. Additionally, global cooperation can facilitate the sharing of knowledge, resources, and technologies, allowing for more efficient and effective solutions to be developed and implemented across the industry.
6. **Monitoring and Evaluation**: Implementing robust mechanisms to monitor the progress of sustainability initiatives and assess their impact on achieving SDGs. This includes conducting regular audits and evaluations to track the performance of fashion companies and measure their contribution toward sustainable development goals (SDGs). By having a systematic monitoring and evaluation process in place, stakeholders can identify areas of improvement, address any shortcomings, and ensure that the industry is making tangible progress toward a more sustainable future. This will enhance transparency and accountability and provide valuable insights to guide future decision-making and policy formulation.

We must request a global call to action, urging policymakers to view sustainable practices in the textile industry as an environmental necessity and an opportunity for economic growth and societal well-being. By implementing an evaluation process, stakeholders can measure the effectiveness of sustainable practices and identify specific areas where improvements can be made. This will enable them to address shortcomings and make necessary adjustments, ultimately leading to a more sustainable future for the textile industry.

Furthermore, the transparency and accountability of this evaluation process will allow for valuable insights to be gained, which can guide future decision-making and policy formulation. It is crucial to urge policymakers worldwide to recognize the potential economic growth and societal well-being that can be achieved by embracing sustainable practices in the textile industry.

References

Alves, L., Cruz, E., Lopes, S., Faria, P., & Cruz, A. (2021). Towards circular economy in the textiles and clothing value chain through blockchain technology and iot: A review. *Waste Management & Research the Journal for a Sustainable Circular Economy, 40*(1), 3–23. https://doi.org/10.1177/0734242x211052858

de Arruda-Filho, N. P., Hino, M. C., & Beuter, B. P. (2019). Including SDGs in the education of globally responsible leaders. *International Journal of Sustainability in Higher Education, 20*(5), 856–870. https://doi.org/10.1108/IJSHE-01-2019-0032

Boiral, O., Henri, J. F., & Talbot, D. (2012). Modeling the impacts of corporate commitment on climate change. *Business Strategy and the Environment, 21*(8), 495–516. https://doi.org/10.1002/bse.723

Boström, M., & Micheletti, M. (2016). Introducing the sustainability challenge of textiles and clothing. *Journal of Consumer Policy, 39*(4), 367–375. https://doi.org/10.1007/s10603-016-9336-6

Chan, T., & Wong, C. W. Y. (2012). The consumption side of sustainable fashion supply chain: Understanding fashion consumer eco-fashion consumption decision. *Journal of Fashion Marketing and Management, 16*(2), 193–215. https://doi.org/10.1108/13612021211222824

Chang, R., & Zhu, S. (2022). Probe into the low-carbon economic environment challenges and opportunities of textile and garment export. *Journal of Environmental and Public Health, 2022*, 1–11. https://doi.org/10.1155/2022/6043903

Chen, L., Caro, F., Corbett, C., & Ding, X. (2019). Estimating the environmental and economic impacts of widespread adoption of potential technology solutions to reduce water use and pollution: Application to china's textile industry. *Environmental Impact Assessment Review, 79*, 106293. https://doi.org/10.1016/j.eiar.2019.106293

Chen, W., Nalluri, V., Hung, H., Chang, M., & Lin, C. (2021). Apply dematel to analyzing key barriers to implementing the circular economy: An application for the textile sector. *Applied Sciences, 11*(8), 3335. https://doi.org/10.3390/app11083335

Cao, D. & Yao, L. (2021). Research on social responsibility risk assessment system of overseas investment of textile and garment enterprises. *Destech Transactions on Economics Business and Management*. https://doi.org/10.12783/dtem/ahem2020/35319

Díaz-Soloaga, P. (2019). The role of communication in organizational culture: Is there a pattern in Spanish fashion companies? *El Profesional de la Información, 28*(5), 1–10. https://doi.org/10.3145/epi.2019.sep.06

Dursun, E., Ulker, Y., & Gunalay, Y. (2022). Blockchain's potential for waste management in textile industry. *Management of Environmental Quality an International Journal, 34*(4), 1174–1197. https://doi.org/10.1108/meq-03-2022-0085
Eccles, R. G., Ioannou, I., & Serafeim, G. (2014). The impact of corporate sustainability on organizational processes and performance. *Management Science, 60*(11), 2835–2857. https://doi.org/10.1287/mnsc.2014.1984
Fadara, T., & Wong, K. (2022). A decision support system for sustainable textile product assessment. *Textile Research Journal, 93*(9–10), 1971–1989. https://doi.org/10.1177/00405175221135167
Fornes, G., Monfort, A., Ilie, C., Koo, C. K. (Tony), & Cardoza, G. (2019). Ethics, responsibility, and sustainability in MBAs. Understanding the motivations for the Incorporation of ERS in less traditional markets. *Sustainability, 11*(24), 7060. https://doi.org/10.3390/su11247060
H&M. (2022). H&M Group Sustainability Disclosure Report 2022. Retrieved from https://hmgroup.com/wp-content/uploads/2023/03/HM-Group-Sustainability-Disclosure-2022.pdf
Islam, M. T., Jahan, R., Jahan, M., Howlader, M. S., Islam, R., Islam, M. M., Hossen, M. S., Kumar, A., & Robin, A. H. (2022). Sustainable textile industry: An overview. *Non-Metallic Material Science, 4*(2), 15–32. https://doi.org/10.30564/nmms.v4i2.4707
Jain, M. (2022). Human resource issues and challenges prevailing in Indian textile & garment industry. *International Journal of Home Science, 8*(3), 227–232. https://doi.org/10.22271/23957476.2022.v8.i3d.1378
Jacometti, V. (2019). Circular economy and waste in the fashion industry. *Laws, 8*(4), 27. https://doi.org/10.3390/laws8040027
Jiang, L., Folmer, H., & Bu, M. (2016). Interaction between output efficiency and environmental efficiency: Evidence from the textile industry in Jiangsu Province, China. *Journal of Cleaner Production, 113*, 123–132.
Kumar, V., Agrawal, T., Wang, L., & Chen, Y. (2017). Contribution of traceability towards attaining sustainability in the textile sector. *Textiles and Clothing Sustainability, 3*(1). https://doi.org/10.1186/s40689-017-0027-8
Li, Y., & Yi, W. (2019). Double decoupling effectiveness of water consumption and wastewater discharge in china's textile industry based on water footprint theory. *PeerJ, 7*, e6937. https://doi.org/10.7717/peerj.6937
Liu, J., Liang, J., Ding, J., et al. (2021). Microfiber pollution: An ongoing major environmental issue related to the sustainable development of textile and clothing industry. *Environment Development and Sustainability, 23*(8), 11240–11256. https://doi.org/10.1007/s10668-020-01173-3
McNeill, L. S., & Moore, R. (2015). Sustainable fashion consumption and the fast fashion conundrum: Fashionable consumers and attitudes to sustainability in clothing choice. *International Journal of Consumer Studies, 39*(3), 212–222. https://doi.org/10.1111/ijcs.12169
Méndez-Suárez, M., Monfort, A., & Gallardo, F. (2020). Sustainable banking: New forms of investing under the umbrella of the 2030 agenda. *Sustainability, 12*(5), 2096. https://doi.org/10.3390/su12052096
Monfort, A., & Mas-Iglesias, J. M. (2021). Barreras y oportunidades para la comunicación de la responsabilidad social en redes sociales. *Comunicación y Hombre, 17*(17), 349–361. https://doi.org/10.32466/eufv-cyh.2021.17.645.349-361
Moorhouse, D., & Moorhouse, D. (2017). Sustainable design: Circular economy in fashion and textiles. *The Design Journal, 20*(1), S1948–S1959. https://doi.org/10.1080/14606925.2017.1352713
Negrete, J. D. C., López, V. N. (2020). A sustainability overview of the supply chain management in textile industry. *International Journal of Trade, Economics and Finance, 11*(5), 92–97. https://doi.org/10.18178/ijtef.2020.11.5.673
Niu, C. & Zhu, Y. (2021). Evaluation index system for green development of textile and garment industry. In *E3s Web of Conferences* (Vol. 275, pp. 02036). https://doi.org/10.1051/e3sconf/202127502036
Oelze, N. (2017). Sustainable supply chain management implementation–enablers and barriers in the textile industry. *Sustainability, 9*(8), 1435. https://doi.org/10.3390/su9081435

Okereke, C., & Russel, D. (2010). Regulatory pressure and competitive dynamics: Carbon management strategies of U.K. energy-intensive companies. *California Management Review, 52*(4), 100–124. https://doi.org/10.1525/cmr.2010.52.4.100

Olofsson, L., & Mark-Herbert, C. (2020). Creating shared values by integrating U.N. Sustainable development goals in corporate communication—The case of apparel retail. *Sustainability, 12*(21), 8806. https://doi.org/10.3390/su12218806

Patora-Wysocka, Z., & Sułkowski, Ł. (2019). Sustainable incremental organizational change—a case of the textile and apparel industry. *Sustainability, 11*(4), 1102. https://doi.org/10.3390/su11041102

Patyal, V. S., Sarma, P. R. S., Modgil, S., Nag, T., & Dennehy, D. (2022). Mapping the links between industry 4.0, circular economy and sustainability: A systematic literature review. *Journal of Enterprise Information Management, 35*(1), 1–35. https://doi.org/10.1108/jeim-05-2021-0197

Pérez, A., Collado, J., & Liu, M. T. (2021). Social and environmental concerns within ethical fashion: General consumer cognitions, attitudes and behaviours. *Journal of Fashion Marketing and Management: An International Journal, 26*(5), 792–812. https://doi.org/10.1108/jfmm-04-2021-0088

Pu, C., Boshan, G., & Chen, X. (2021). Introduction of environmental protection concept in textile major education. *Journal of Educational Theory and Management, 5*(1), 60. https://doi.org/10.26549/jetm.v5i1.6471

Qingfen, Z., Wei, G., & Yuwen, Z. (2023). Temporal and spatial trajectories of the economic and pollution emission focus of China's textile industry. *World Journal of Management Science and Technology, 1*(1), 41–51. https://doi.org/10.57237/j.wjmst.2022.01.006

Sadeka, A. & Hammash, A. (2023). The environmental and health impact of eco-friendly textiles in the interior space. *SVU-International Journal of Engineering Sciences and Applications, 4*(1), 35–40. https://doi.org/10.21608/svusrc.2022.159112.1069

Saini, K. (2023). Eggshell-lignin carbon composite for methyl orange removal from wastewater. *Industrial & Engineering Chemistry Research, 62*(38), 15499–15510. https://doi.org/10.1021/acs.iecr.3c02029

Schneider, S., & Clauß, T. (2020). Business models for sustainability: Choices and consequences. *Organization and Environment, 33*(3), 384–407. https://doi.org/10.1177/1086026619854217

Segal, M. (2022). H&M commits nearly $300 million annually to tackle supply chain emissions. *ESGToday.com.* (14 November 2022). Retrieved from https://www.esgtoday.com/hm-commits-nearly-300-million-annually-to-tackle-value-chain-emissions/

Shamsuzzaman, M., Islam, M., Hasan, H., Khan, A., & Sayem, A. (2023). Mapping environmental sustainability of knitted textile production facilities. *Journal of Cleaner Production, 405*, 136900. https://doi.org/10.1016/j.jclepro.2023.136900

Shen, B., Li, Q., Dong, C., & Perry, P. (2017). Sustainability issues in textile and apparel supply chains. *Sustainability, 9*(9), 1592. https://doi.org/10.3390/su9091592

Shehri, F., Rehman, T., Mansoor, M., & Saif, O. (2022). Exploring the role of entrepreneurial leadership in decent work and economic growth: a sustainable development of textile industry of Pakistan. *Journal of Social Sciences Review, 2*(4), 237–248. https://doi.org/10.54183/jssr.v2i4.82

Siedschlag, I., Meneto, S., & Koecklin, M. (2019). Determinants of green innovations: Firm-level evidence. *SSRN Electronic Journal.* https://doi.org/10.2139/ssrn.3495884

Sun, X., Xi, W., Sun, F., Tian, M., Qu, L., Perry, P., Owens, H., & Liu, X. (2021). Textile waste fiber regeneration via a green chemistry approach: a molecular strategy for sustainable fashion. *Advanced Materials, 33*(48). https://doi.org/10.1002/adma.202105174

Tian, W., Huang, K., Zhu, C., et al. (2022). Recent progress in biobased synthetic textile fibers. Frontiers in Materials, 9. https://doi.org/10.3389/fmats.2022.1098590

Vijeyarasa, R., & Liu, M. (2021). Fast fashion for 2030: Using the pattern of the sustainable development goals (sdgs) to cut a more gender-just fashion sector. *Business and Human Rights Journal, 7*(1), 45–66. https://doi.org/10.1017/bhj.2021.29

Wang, L., Li, Y., & He, W. (2017). The energy footprint of china's textile industry: Perspectives from decoupling and decomposition analysis. *Energies, 10*(10), 1461. https://doi.org/10.3390/en10101461

Wood, J., & Redfern, J. (2022). Developing textile sustainability education in the curriculum: Pedagogical approaches to material innovation in fashion. *International Journal of Fashion Design Technology and Education, 16*(2), 141–151. https://doi.org/10.1080/17543266.2022.2131913

Xu, J., & Wang, B. (2019). Intellectual capital performance of the textile industry in emerging markets: A comparison with China and South Korea. *Sustainability, 11*(8), 2354. https://doi.org/10.3390/su11082354

Yurdakul, M., & Kazan, H. (2020). Effects of eco-innovation on economic and environmental performance: Evidence from turkey's manufacturing companies. *Sustainability, 12*(8), 3167. https://doi.org/10.3390/su12083167

Zhang, Z., Cheng, H., & Yu, Y. (2020). Relationships among government funding, r&d model and innovation performance: A study on the Chinese textile industry. *Sustainability, 12*(2), 644. https://doi.org/10.3390/su12020644

Zhu, A., & Hu, Y. (2016). The low-carbon and environmental protection of textile and garment in china. *Journal of Geoscience and Environment Protection, 04*(11), 45–49. https://doi.org/10.4236/gep.2016.411005

Chapter 3
Sustainable Fashion Communication: How to Overcome the Hidden Fear of Spanish Brands

Paloma Díaz-Soloaga, Sonia Carcelén-García, Mónica Díaz-Bustamante, and Olga Kolotouchkina

3.1 Introduction

Public scrutiny of companies since the beginning of the twenty-first century has made it easier for society to discover how they manufacture and distribute their products and has raised social awareness in relation to issues that had not been considered in much depth in the past (Harlow et al., 2013; Mukendi et al., 2020). At the same time, the media discourse on the human impact on the environment has led to a widespread belief that change is possible through the efforts of many individuals (Gestal, 2021; Henninger et al., 2021).

This influence is particularly noticeable in the way the concept of "sustainable" is communicated by companies in all areas: from the food sector to the automotive industry, and from real estate to pharmaceuticals (Adams, 2022; Arthur, 2023; Ellen MacArthur Foundation, 2021; Genç, 2017; Guillen Mandujano et al., 2021). In the discourses of brands, the responsibilities of institutions and citizens are intertwined interchangeably (Woodside and Fine, 2019). There are frequent messages

P. Díaz-Soloaga (✉)
UCM Brand and Integrated Communication Research Group, Complutense University, Madrid, Spain
e-mail: pdiaz@ccinf.ucm.es

S. Carcelén-García · M. Díaz-Bustamante · O. Kolotouchkina
Complutense University, Madrid, Spain
e-mail: sonialca@ucm.es

M. Díaz-Bustamante
e-mail: mdiaz-bu@ucm.es

O. Kolotouchkina
e-mail: olga.kolotouchkina@ucm.es

J. F. Gallego-Nicholls and A. Carrilero-Castillo (eds.), *Textile Industry and the SDGs*, SDGs and Textiles, https://doi.org/10.1007/978-981-97-5073-3_3

from corporations encouraging people, as consumers, to share in the effort to achieve the common goal of sustainability (Harlow et al., 2013; Newig et al., 2013).

The Spanish fashion sector, made up of 19,676 textile, clothing and footwear companies, the vast majority of which are small and medium-sized enterprises, has undergone a notable transformation over the last two decades. The need to internationalise the business due to the financial crisis of 2007, associated with that to digitalize it today, has pushed brands toward different dynamics from those of the past. In particular, the catastrophe that occurred in the collapse of the Rana Plaza building in Bangladesh in 2013, where more than a thousand people died, had a strong effect on the entire industry (Salvá, 2021). From that date onwards, the use of sustainability as a sales argument by fashion brands to encourage their customers to buy began to become normalized (Fischer et al., 2016; Genç, 2017; Woodside and Fine, 2019; Godemann, 2021).

The attributes "natural", "ecological, "environmental", "sustainable", "green", "circular", "committed", "responsible", "ethical" and many others are intended to move consumers' motivations toward conscientious behavior, and to prevent negative perceptions from holding back their purchasing decisions (Vehmas et al., 2018; Kim & Kim, 2020; Capgemini Research Institute, 2021; Díaz-Soloaga, 2021; Arthur, 2023). More and more consumers weigh these adjectives when deciding to purchase products (Chu et al., 2020), and numerous fashion brands have decided to make sustainability their main theme when communicating with their customers (Guillen Mandujano et al., 2021; Becker-Olsen & Potucek, 2013; Córdoba Cubero, 2021; Golob et al., 2023; Adamkiewicz et al., 2022).

But not all consumers are aware that their purchasing behavior can impact the reality behind an item of clothing, influencing other people's working conditions, the environmental balance or the disposal and recycling process (De Jong et al., 2020; Han et al., 2017). In fact, research shows that the average consumer in Spain does not take into consideration the socially responsible commitment of fashion companies when making their purchasing decision (Kantar Worldpanel, 2022; CIS, 2022). This does not mean that they are not concerned about sustainability issues, but it is an aspect that they associate more directly with other industries, such as electricity generating, the automotive sector, and even food, rather than with fashion and textiles (Martínez, 2015; Rausch & Kopplin, 2021).

It is in this context that inaccuracies, exaggerations and, on occasions, deception arise in the communication of sustainability by many fashion companies. Consumers tend to distrust the messages offered by companies if they do not have enough information to verify them (Alejo Góngora, 2013; Becker-Olsen & Potucek, 2013). This practice is known as *greenwashing*, a term that is still largely unknown to a large part of the population (Roger & Buil, 2014; European Union, 2021; European Commission, 2021).

It is such an important issue that public institutions, both national and international, and multinational corporations are trying to regulate the use of language in sustainability communications in order to avoid the improper or abusive use of these terms (BOF & McKinsey, 2022). In this way, the aim is to achieve a better match between the reality companies' practices and the way they tell their different stakeholders about the sustainable commitments they are undertaking, without exaggerating, generalizing and much less, lying about them.

The European Union has recently regulated these contents through directives that each country must implement within a certain period (European Commission, 2023). As sustainability is a social value that is widely accepted by society, the intention is to protect consumers from deception and manipulation.

The control of erroneous communication practices can be of great help to companies, if in the end they end up curbing the phenomenon of *greenwashing*, since there are also many positive actions that can be highlighted by corporations and in this way reinforce their reputations. The efforts that many Spanish brands have made over the last decade to improve their so-called responsible "behavior indicators" must be considered positively. These include commitments to reduce CO2 emissions, careful selection of the fibers they use, obtaining certificates and quality seals at different stages of the production chain, improvements in circularity (management of factory surpluses, reuse of garments, reverse logistics, etc.

Finally, it should be stressed that there are companies in the Spanish fashion sector that do not consider it relevant to build their narrative on sustainability and yet follow a business model that is absolutely compatible with this concept. They are companies that reinforce the value of tradition and heritage through the origin of their shapes and patterns, artisan labor, the slow consumption they promote or the natural fiber they use. In this way, they develop values related to and compatible with sustainability, without the need to expressly mention them.

3.2 What Do Consumers Think and What Do Brands Say?

During the years 2022 and 2023, some members[1] of the Brand Management Research Group of the Complutense University of Madrid, carried out an ambitious research project to gain in-depth knowledge of the Spanish sustainable fashion consumer and their attitude toward the commercial communications developed by Spanish brands. The aim was also to identify the narratives used by fashion companies on social networks to build their brand story.

We structured the project around two phases, which were carried out in parallel by the five members of the research team. Therefore, we decided to develop a consumer study, obtaining data through a survey of Spanish consumers, and also a content analysis of the Instagram profiles of a sample of Spanish fashion brands representative of different consumer segments was undertaken.

[1] This project was carried out thanks to funding from Santander–UCM. Along with the research team, led by Paloma Díaz Soloaga, technical assistance was provided by Paula Osuna Maté.

3.3 First Phase of the Study

To understand the extent to which *greenwashing* is an activity that brands can avoid through truthful and effective communication, the first objective of this project was to identify the perceptions, attitudes and behaviors of Spanish consumers in relation to sustainable fashion in order to draw up a state of play from a consumer perspective. A second objective was to assess consumers' interest in the sustainability content of Spanish fashion brands. Finally, we wanted to find out whether consumers interpret sustainability communications from fashion brands as *greenwashing*, as they do not believe them to be entirely reliable information (Szabo & Webster, 2021) (Fig. 3.1).

This first part of the study was carried out in November 2022 using a survey of 1,000 people representative of the Spanish population in terms of their basic socio-demographic characteristics such as age, sex, province, level of education and occupation (Table 3.1).

In addition to the above questions, participants also assessed in the questionnaire their perception of the sustainability of 21 Spanish fashion brands (Zara, Mango, Ecoalf, Adolfo Domínguez, Berska, Bimba y Lola, Cortefiel, Desigual, El Corte Inglés, El Ganso, Scalpers, Stradivarius, Sepiia, SKFK (Skunfunk), Pepe Jeans, Pedro del Hierro, Loewe, Purificación García, Roberto Verino, Massimo Dutti and Springfield). With the objective of understanding the consumer and subsequently, in a second phase of the project, addressing the content of messages issued by brands, more questions related to the purchase of fashion products, the consumption of information about sustainable fashion, the actions taken the sustainability policy communicated by brands and the perception of greenwashing in fashion brands.

The distribution of the sample corresponds to the structure of the resident population in Spain by gender, age and Autonomous Community of residence (INE Base, 2022). This sample size has an indicative error of ±3.15% for a confidence level of 95.5% (P = Q = 50% and 2 sigma).

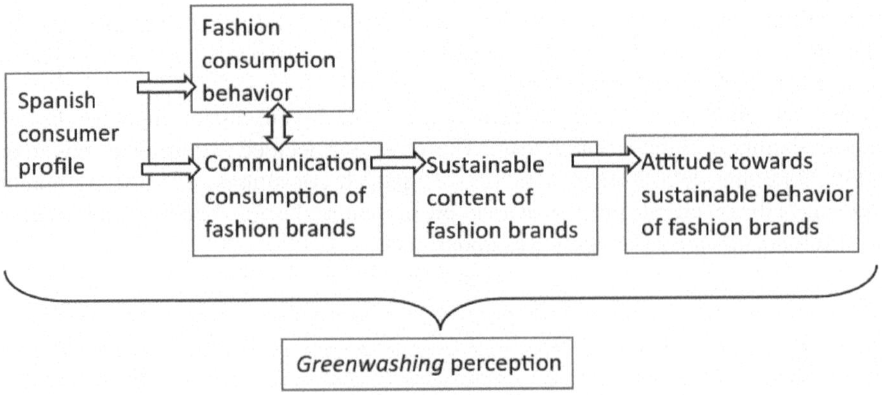

Fig. 3.1 The dimensions studied to understand consumer perceptions of greenwashing. *Source* Authors

Table 3.1 Elements studied and variables used

Table elements	Variables
1. Socio-demographic characteristics	Sex Age Level of education Occupation
2. Sustainable fashion purchasing behavior	Attributes associated with corporate social responsibility Environmental type attributes Attributes related to the marketing mix
3. Sustainable fashion content consumption behaviors	You are often looking for information on sustainable fashion-related content
	Use of different channels to seek or receive information on sustainable fashion Consumer interest in sustainable fashion content
4. Attitudes toward sustainability policies as communicated by fashion -companies	Sustainability is used by some companies to launder or improve their corporate image or reputation I am not interested in sustainability information from fashion brands
5. Perception of *greenwashing* in the studied fashion brands	The brand uses sustainable fashion messages to wash or improve its image
6. Level of sustainability perceived in Spanish brands	How sustainable each brand is considered

Source Authors

3.4 Results on Attitudes and Behaviors Towards Sustainable Fashion

An initial result of interest was the associations that consumers (women and men) relate in the first instance to the concept of sustainability. At the same time, these attributes break down, inversely, the interests that consumers show when they try to bring the abstract idea of "sustainability" down to earth. It is interesting to note that they all prioritize environmental balance over other aspects such as workers' working conditions and brands' social commitment. Of course, design and price, as usual in these studies, always come last (Fig. 3.2).

Secondly, when investigating consumers' interest in information on sustainability actions carried out by fashion brands, it was found that only a third of the consumers surveyed (32.4%) actively seek out this type of content, which shows that this information is still not relevant for a significant part of society, for reasons that the following graph helps to better understand (Table 3.3).

The topic of greatest interest to respondents is what fashion products are made of, followed by the environment. This factor is possibly a consequence of the communication work that brands have done to raise awareness of the origin of fibers and

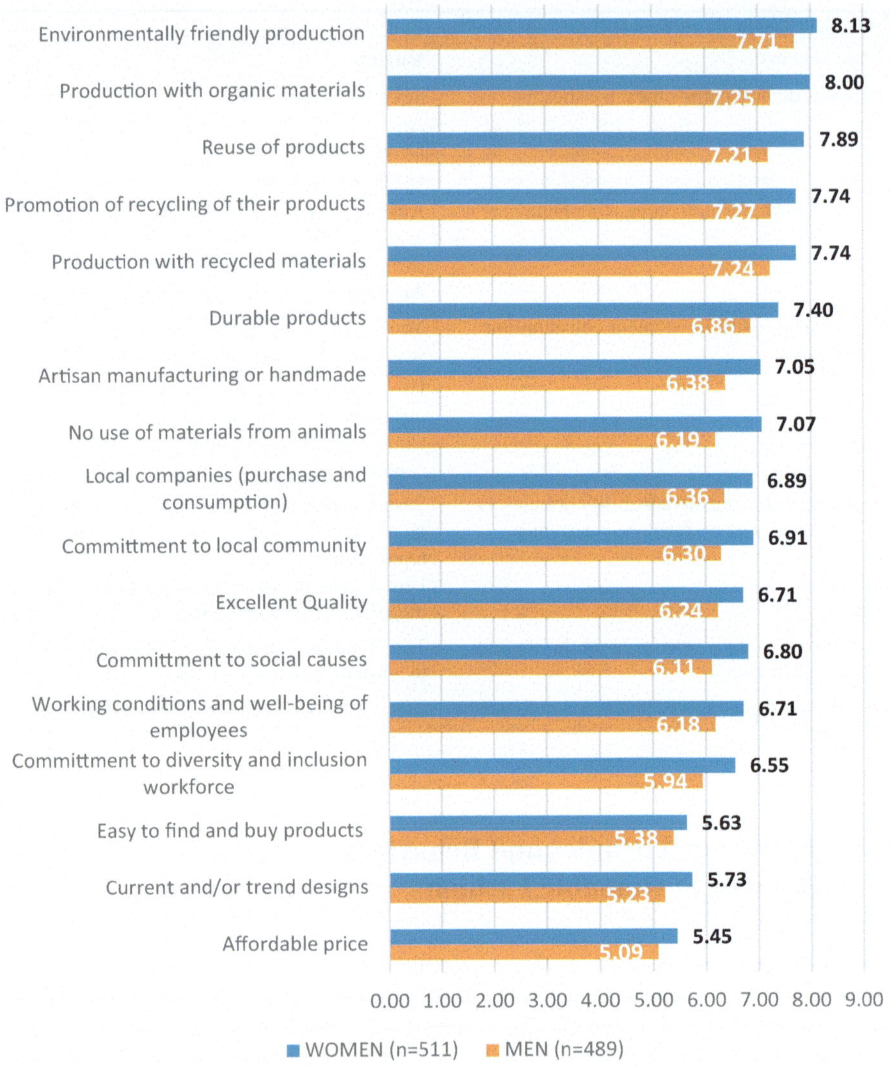

Fig. 3.2 Attributes associated with sustainable fashion*. *Source* Authors. *Total sample (n = 1,000). Data in averages on a scale of 0 to 10)

their production processes. Organic cotton, silk, linen, wool labels are the most recognizable distinguishing element of fashion brands.

Along with this, the other striking fact is the lower rating given to factors relating to social and labor issues in fashion companies. Thus, manufacturing conditions, social projects and the composition and circumstances of the workforce are topics of less interest to consumers. The same applies to messages calling for responsible consumption. It remains to be studied how consumers interpret these types of narratives when they originate from a fashion company (Fig. 3.3).

When looking at the perception of sustainability of certain brands compared to others, the result gives priority to those that have carried out a constant communication exercise focused on sustainability over long periods, as is the case of Ecoalf, Adolfo Domínguez, SKFK and Sepiia. All these brands define themselves in their communication channels (their website and social media profiles) as expressly committed to the values of sustainability, and they do so while carrying out their business of distributing fashion and trends each season.

Although this is an analysis of respondents' perceptions, it is worth highlighting the clear association between the business model's entrepreneurial commitments and

Fig. 3.3 Consumer interest in sustainable fashion content*. *Source* Authors. *Total sample (n = 1,000). Data in means on a scale from 0 to 5)

the brands' communication strategies. An association that has proven to be effective (Fig. 3.4).

Another relevant aspect is again the gender difference found in some of the opinions given by consumers. Thus, women stand out above men, both in the purchase of sustainable fashion products and in the search for content related to the subject. The fact of being a public with a greater availability toward fashion at the same time gives them a special sensitivity toward aspects related to sustainability (Fig. 3.5).

The study shows that Spanish consumers are quite critical of the use of sustainability as a main theme in the commercial communication of fashion brands. So much so that 71.9% of the sample analyzed considered that fashion brands that use *greenwashing* as a way of improving their corporate reputation. Fifty eight point five percent believe that sustainability is a good sales argument used by brands, but quite

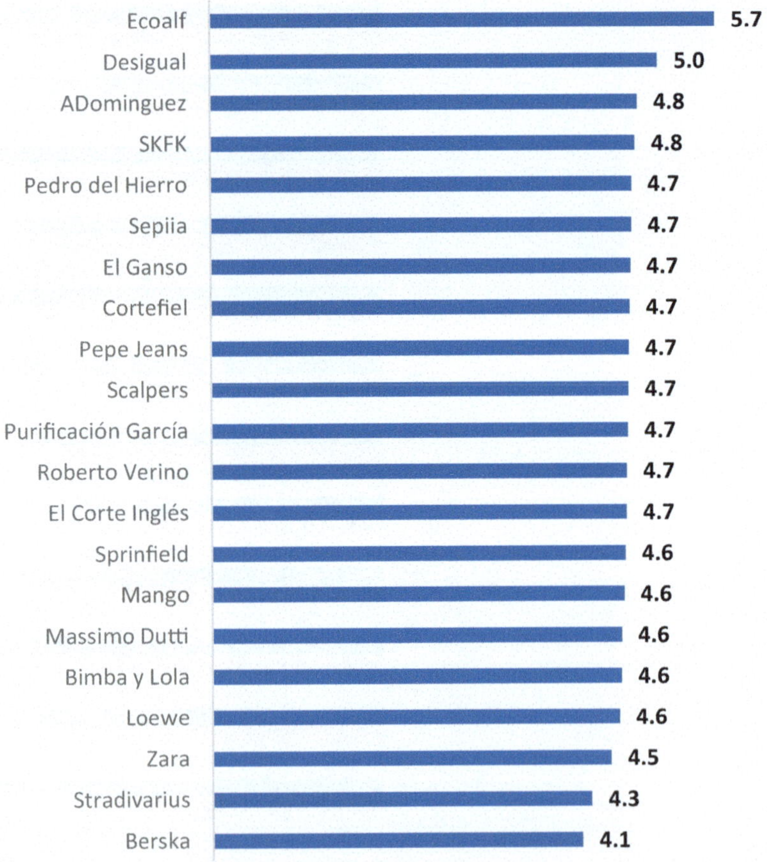

Fig. 3.4 Level of sustainability perceived in Spanish brands*. *Source* Authors. *Total sample (n = 1,000). Data in means on a scale from 0 to 5)

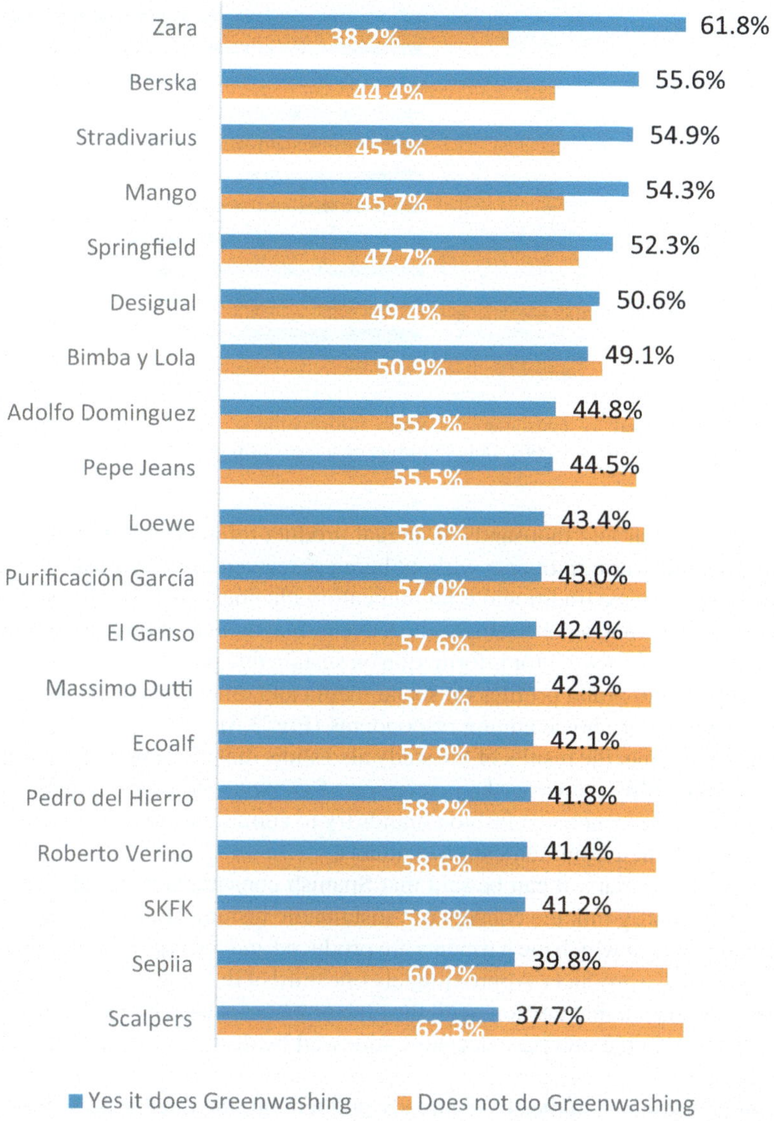

Fig. 3.5 Perception of *greenwashing* in Spanish brands*. *Source* Authors. *Total sample (n = 1,000). Data in means on a scale of 0 to 5)

a few consumers (62%) believe that this argument is not backed up by real sustainable action. Some of the best-known brands with the highest turnover in Spain were identified as the brands with the highest perception of *greenwashing*. In this regard, more than half of the respondents (53.7%) doubt that fast fashion brands, with several

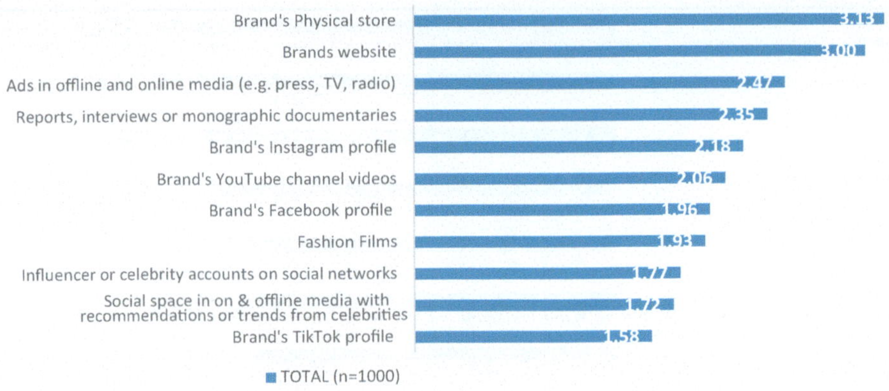

Fig. 3.6 Channels used by consumers to obtain information on sustainable fashion*. *Source* Authors. *Total sample (n = 1,000). Data in means on a scale of 0–5)

collections per year and thousands of annual product references, can communicate their sustainability commitments without being perceived as *greenwashers*.

Interestingly, more traditional communication channels such as physical shops, corporate websites and advertisements in conventional media continue to be preferred by customers when looking for information on sustainable policies of fashion companies, as opposed to brand profiles on social media and other digital channels, which are the least popular choice among respondents (Fig. 3.6).

At the same time, the results of the study also show that actively seeking information on sustainability increases the perception of *greenwashing* among respondents. This ratio increases when fast fashion brands try to communicate their sustainability actions through their communication channels.

By way of summary, it can be said that Spanish consumers (through the sample analyzed), primarily link the concept of "sustainable fashion" to caring for the environment, the way in which the garments are produced and the type of material used to manufacture these products. Approximately one third of the sample (36.8%) declared that they buy sustainable fashion, with the type of material used being the most influential attributes in the purchase decision, followed by design, price and information in the shop.

As other academic studies have already pointed out, the three aspects least associated with the idea of sustainable fashion continue to be the difficulty of accessing or buying this type of fashion, the lack of a current or trendy design, and the high price. The three Spanish brands perceived as the most sustainable by those surveyed were: Ecoalf, Desigual and Adolfo Domínguez (Fig. 3.4).

Another major conclusion drawn from the descriptive analysis of the study is that greenwashing is a major threat to the fashion industry, as it is a fairly deep-rooted consumer perception (71.9% of respondents agree that sustainability is used by some companies to wash or improve their image). A somewhat more detailed analysis of the data indicates that this perception increases up to 79.3% for those consumers who declare that they are not interested in information on sustainability as a sales argument for brands.

This finding needs to be explored further, as it is of great interest for the future of fast fashion brands, as well as being a positive brand asset for those that have been communicating sustainability for some time.

Contrary to what it may seem at first glance, the sustainable communication actions carried out by brands (either through traditional advertising campaigns or recommendation using *influencers*), do not seem to have a great impact on consumers when purchasing a sustainable fashion garment, which can also be useful to know for brands that are working on their positioning in these brand values.

3.5 Second Phase of the Study

In order to get to know the lines of argument and resources used by brands when communicating, it was decided to focus the study on the Instagram social network, as it is the most used by brands, along with their own websites, to transmit their identity and ethos.

In this second phase, the aim was to delve deeper into the main communication strategies and narratives used by Spanish fashion brands in their profiles on social networks and digital channels, on the understanding that they are a privileged place for brands to develop their sustainable story (Alonso González, 2015; Milanesi et al., 2022; Pérez-Curiel et al., 2021; Shen, 2021). For this purpose, 49 Spanish brands were selected (Table 3.2), with a turnover of more than one million euros and an Instagram profile. After a first exploratory stage of the brands carried out with the Metricool software, the five brands that most frequently used sustainability arguments in their Instagram *profile posts and reels* were selected: Adolfo Domínguez, Mayoral, Ecoalf, SKFK and Boboli. It is worth noting that most of the 49 brands initially studied never (34.6%) or almost never (46.9%) referred to the concept of sustainability in any of its forms and associations during the time investigated.

It was decided to analyze the content of these brands for one month (15 September–15 October 2022) using the *4 k Stogram* tool, which allows downloading the content (*likes*, *hashtags*, stories, conversation) of the brands.

3.6 Narrative Outcomes and Discourses Present on the Instagram Social Network

An investigation of the most frequently used discourses when discussing sustainability in Instagram posts shows the frequency of each of them (Fig. 3.2).

As can be seen in Fig. 3.7, the most common discourse is the "use of organic materials", which is present in a quarter of the publications analyzed, with a slightly

Table 3.2 Sample of the brands analyzed in the study

Adolfo Domínguez	Mango	Rosa Clara
Berska	Maria Pascual	Scalpers
Bimba y Lola	Massimo Dutti	Sepiia
Boboli	Mayoral	SKFK
Brownie	Muroexe	Sophie and Lucie
Camper	Name the brand	Springfield
Castañer	Neck & Neck	Stradivarius
Cortefiel	Oysho	System Action
Decimas	Pedro del Hierro	Trasluz
Desigual	Pedro García	Tuc
Ecoalf	Pepe Jeans S.L	Unisa
El Corte Ingles	Pikolinos	Vilagallo
El Ganso	Pitillos	We Are Knitters
Gocco	Pompeii	Women Secrets
HOFF	Pronovias	Zara
Hoss Intropia	Purificación García	
Loewe	Roberto Verino	

Source Authors

Table 3.3 Frequency of inclusion of sustainability content

	Adolfo Domínguez	Boboli	Ecoalf	Mayoral	SKFK
Inclusion percentage	28,12	23,07	89,65	5,5	37,93
Engagement ratio	1,42	1,14	01,04	01,02	2,36

Source Authors

lower presence of the "recycling of products" discourse. In third and fourth positions are "allusion to the environment" and "responsible consumption", respectively. There was a lower presence of discourses related to "care for the workforce", "use of clothing" and "energy saving".

These aspects that are less addressed in communication strategies are equally relevant in the area of sustainability, indicating that there is still room for improvement in their integration into brand communication strategies.

As sustainability is a broad and multidimensional concept that encompasses different facets, such as environmental, social and economic aspects, we realized that the brands analyzed emphasise the impact on the environment in particular, neglecting aspects related to the manufacture and production of garments, as well as other variables such as the cost and impact of transport, reverse logistics, or the subsequent impact of garments when it comes to recycling or the use of leftover textile products. It is strongly recommended that future strategies include aspects related to work ethics, social justice and a more responsible management of the

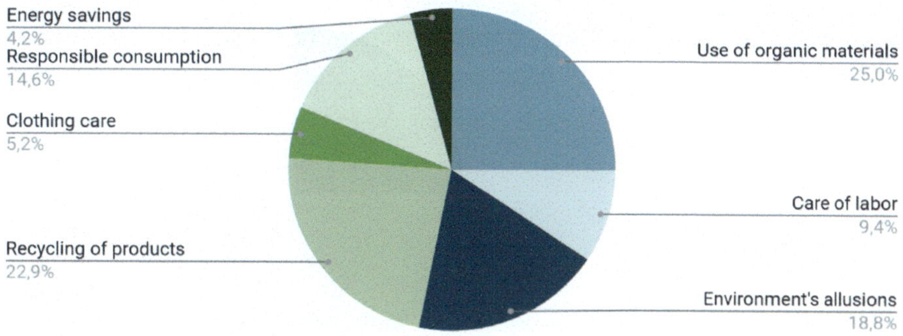

Fig. 3.7 Fashion brands' main arguments for communicating sustainability. Source: Authors

resources provided by people; this recommendation will be further developed in the conclusions.

The next question analyzed relates to the recommended frequency of sustainability content publishing for fashion brands. From the beginning of the project, the question was raised as to what should be the appropriate volume of sustainability publications compared to the rest of the brand's organic content. In order to obtain the most effective combination, a relationship has been established between the different percentages of inclusion of each brand and the average engagement rate. In this way, the aim was to determine the ideal frequency that generates the greatest impact and participation from the audience. Table 3.3 shows the different values for each of the aforementioned variables.

When considering *engagement*, we see that Adolfo Domínguez and SKFK obtained the best results, in addition, in both cases they reached inclusion percentages of around 30% and 40%, which helps to determine the relative weight of content related to sustainability compared to other brand messages (such as product, inspiration or consumer warnings, to name a few topics).

On the other hand, it is found that the brands at the extremes, i.e., those with the highest and lowest percentage of inclusion, are the worst performers in terms of *engagement*. This may indicate that there is a balance point in the frequency of publishing sustainable content that is more effective in generating audience interest and engagement.

The third question was to determine which formats are most effective in communicating sustainability messages. According to the set of data obtained, it can be seen that the format most used by the brands analyzed is the static post, representing approximately 75% of the total number of publications. The use of *reels* was less frequent, representing only 25% of the occasions. However, it is noteworthy that three of the five brands studied have chosen to use a combination of formats in their content strategies. This suggests that the diversification of formats may be an interesting strategy to capture the attention of the audience and generate a greater impact in terms of *engagement* (Fig. 3.8).

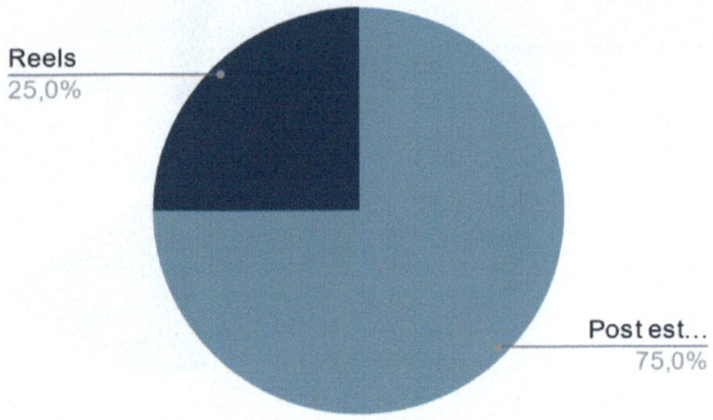

Fig. 3.8 Distribution of formats. *Source* Authors

Next, another aspect that we sought to clarify concerning the formats used is whether there is evidence that one obtains better results than the other. To this end, a comparison was made between the average *engagement* of the *reel* format and the static post. In this case, it was observed that both formats have a very similar *engagement* index, with the static post being slightly higher, with 0.03 more. This indicates that it cannot be categorically stated that one format is more effective than the other in terms of audience interaction and participation. It is important to bear in mind that the success of a format may depend on several factors, such as the creativity of the content, the relevance of the message and the emotional connection it establishes with the followers. Therefore, it is advisable to consider a combination of formats in a content strategy to make the most of the possibilities that each one offers (Fig. 3.9).

Table 3.4 was then drawn up to identify the implementation of certain techniques (use of hashtags, emoticons, narrative structures, use of responses or links) in the brands' communication strategies.

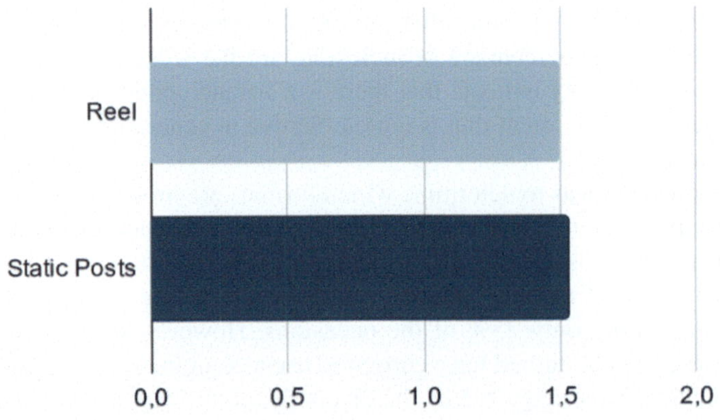

Fig. 3.9 Engagement results for each format. *Source* Authors

Table 3.4 Use of resources to increase outreach

BRAND	Hashtags	Emoticons	Interactive structure	Answers	Links
A. Domínguez	Yes	Yes	No	Yes	No
Boboli	Yes	Yes	Yes	No	No
Ecoalf	Yes	Yes	Yes	Yes	No
Mayoral	Yes	Yes	Yes	No	No
SKFK	No	Yes	Yes	No	No

Source Authors

When analyzing the results, it was observed that the frequent use of hashtags is a fairly common tagging strategy in sustainability-related publications, with the surprising exception of the Basque brand SKFK.

Brands use hashtags related to their identity and purpose, as well as terms directly related to sustainability. Therefore, it can be inferred that their strategic use can be beneficial to increase visibility and engagement with sustainability-related posts.

The widespread use of emoticons in the brands analyzed also stands out. The presence of these elements in publications related to sustainability has been a recurring strategy in every one of the cases studied. These symbols give dynamism to the publications and increase their expressive potential, which is attractive to the audience.

Another key to maximizing the average reach of publications is the use of interactive structures. In this regard, four of the five brands studied have used this practice in their sustainability content. Interactive structures can take different forms, such as questions, surveys or invitations to share their own experience. These strategies seek to engage the audience and build an active community around the interest in sustainability.

In relation to the above, only Adolfo Domínguez and Ecoalf follow the practice of responding to questions and comments from their followers, believing it to be a two-way strategy that is highly valued by their users. It is important to note that the possibility of interacting with followers is conditioned by the presence of comments on posts. In many cases, the publications did not generate this type of interaction, which reduced the brands' opportunities to respond in real time to their followers' questions and queries.

The last interaction technique analyzed is the incorporation of links in publications. The use of this practice makes it possible to redirect user traffic to other brand websites. In this case, none of the brands in the study have chosen to incorporate links to their "permanent" posts, i.e., those that remain in the profile *feed.* However, it is necessary to remember that this research is based solely on the analysis of posts in the *feed*, and we cannot rule out the possibility that brands have implemented this strategy in other types of content, such as stories, where it is more common to find links.

In summary, it is concluded that the strategic use of hashtags related to brand and sustainability, as well as the inclusion of emoticons in posts, are effective practices to expand reach and generate interest in sustainability content. In addition, implementing interactive structures promotes audience participation and engagement with posts. While some brands have responded to follower questions and interactions, this practice has been limited.

3.7 Conclusions and Recommendations

Due to the interest in sustainability in society and the public scrutiny of institutions, more and more companies are questioning their corporate values, thus initiating a paradigm shift (Weder et al., 2021).

Thirty years after the definition of the TBL (Triple Bottom Line) model by John Elkington (1994), *People, Planet and Profit, the* development of policies that include sustainable approaches is associated with an increase in value for the company and this not only translates into an increase in economic performance, but also the perception of a "committed brand". It also becomes a mark of differentiation and a long-term risk prevention barrier. Committing to sustainable and transparent strategies will improve consumers' trust and perception of the brand (Vehmas et al., 2018).

This study highlights the fact that the narrative of "sustainability in fashion" encompasses ecological, social and ethical aspects, which ensure conscious and responsible human and corporate development.

Spanish fashion companies focus primarily on the first dimension (the environmental issue), paying much less attention to the human dimension (fair working conditions, health and well-being of the professionals involved) and social aspects (irresponsible consumption, possible consumer addictions). It would therefore be highly advisable for brands to elaborate more equitable messages on these dimensions in order to better balance the concept of sustainability.

At the same time, the starting point of this chapter was precisely to analyze the paradox that occurs in the process of communicating sustainability in the fashion industry. Large Spanish fashion companies perceive that the more they make an effort to communicate their sustainable processes and actions, the more easily they are accused of *greenwashing* (Martín López, 2019). It is as if the consumer replies to brands with the Latin phrase *"excusatio non petita accusatio manifesta"*. If they talk so much about how they do things in the field of sustainability, is it not because they are hiding something?

Another cause could be the criticism directed at the business model of so-called "fast fashion" itself, which is not only identified with fast production and distribution but at the same time is associated with the encouragement of intensive, accelerated and constant consumption of fashion.

At the same time, it is important to consider the effort and real commitment of these brands to carry out their industrial processes respecting production standards that are often better than the regulations of each country. Inditex, Mango, Desigual, Tendam, Sociedad Textil Lonia, Mayoral, Adolfo Domínguez, El Corte Inglés, to name but a few, are companies that closely monitor their suppliers, the composition of their raw materials, the quality of their employees' work and other dimensions of the broad concept of sustainability.

Similarly, when implementing digital communication strategies, brands must find the right balance between communicating their commitment to sustainability and avoiding saturating consumers with too much information. Regular inclusion of sustainability-related content helps to maintain consumer interest and awareness, without losing their attention due to overexposure. Using quality certifications (such as B corp, GOTS, Oeko-tex, OCS, among others) and external evaluations or audits is another resource used by some of the brands analyzed and has proven to be an effective action, providing security to consumers who are unable to verify for themselves the veracity of the claims they make.

Instead of approaching sustainability communication as a compulsory task, brands should see it as a great opportunity to generate higher-quality content that engages their followers. To do this, it is crucial to understand the characteristics of each channel and explore the possibilities of different formats to achieve creative storytelling. This effectively conveys sustainability content and captures attention in an innovative way.

The examination of the strategies implemented by the different brands reveals an interest in creating a conversation around sustainability. The study showed how the SKFK brand shares its history of transformation throughout its existence, as well as all the actions it is currently undertaking to contribute to a better world, through the communication of the sustainable commitments it makes in several dimensions: certificates that guarantee the commitment to ensure fair relations with all the people in their production chain, as well as the cultivation of organic cotton in an Indian community, and garment renting as a proposal for conscious consumption. As the business is convinced of the rightness of this path the aesthetics used transmit a sensation of harmony and calm, often using images of nature, warm colors and smiling female figures in relaxed attitudes dressed in their garments.

Mayoral and Boboli, two children's fashion brands, are a good example of communicating simple and understandable messages for a consumer who is not very experienced with sustainable narratives. Their discourses are based on the reality of their commitments and processes, mainly related to the environment.

Adolfo Domínguez has been sensing for decades the relevance of working with technological fabrics that reduce the impact of industry on the animal world. Their commitment to textile research has led them to great achievements in this field. But their most courageous stance has been to fight for more responsible fashion consumption. With campaigns such as "Be older", it has opted for *Slow Fashion*, which advocates the value of quality pieces that last for decades in the wardrobe of its consumers and do not age. In both cases, messages are launched that appeal to the consumer, trying to awaken their attention and then move them to a greater commitment.

Ecoalf is the paradigm of brand positioning, having managed in its short existence to occupy a space in sustainable fashion that has recycling, design and technology at a reasonable price as its communicative axes. To achieve this, it has been essential to make consistent and constant use of a shared narrative with an international consumer interested in creative solutions to the problem of environmentally polluting waste.

The five cases analyzed show that we are witnessing the origin of a way of communicating sustainability linked to their identity and with a strong connection with audiences. They are brands that choose sustainability as a fundamental axis of their narrative and demonstrate that they not only manage to improve their digital results, but also position themselves as leaders and benchmarks in the adoption of sustainable practices.

References

Adamkiewicz, J., Kochanska, E., Adamkiewicz, I., & Łukasik, R. M. (2022). Greenwashing and sustainable fashion industry. *Current Opinion in Green and Sustainable Chemistry, 38*, 100710.

Adams, C. A. (2022). Sostenibilidad y la empresa del futuro, BBVA.

Alejo Góngora, C. L. (2013). Greenwashing: Ser verde o parecerlo. Cátedra La Caixa Responsabilidad Social Corporativa. IESE Business School.

Alonso González, M. (2015). Las redes sociales como canal de comunicación de las marcas de moda españolas: El caso Zara, Mango y El Corte Inglés. *Revista INDEX Comunicación, 5*, 77–105.

Arthur, R. (2023). The sustainable Fashion Communication playbook. Shifting the narrative: A guide to aligning fashion communication to the 1.5-degree climate target and wider sustainability goals. UN environment project. Fashion for climate.

Becker-Olsen, K., & Potucek, S. (2013). Greenwashing. In S. O. Idowu, N. Capaldi, L. Zu, & A. D. Gupta (Eds.), *Encyclopedia of corporate social responsibility*. Springer. https://doi.org/10.1007/978-3-642-28036-8_104

BOF & McKinsey. (2022). The State of Fashion 2023.

Capgemini Research Institute. (2021). How sustainability is fundamentally changing consumer preferences.

Chu, S. C., Chen, H. T., & Gan, C. (2020). Consumers' engagement with corporate social responsibility (CSR) communication in social media: Evidence from China and the United States. *Journal of Business Research, 110*, 260–271. https://doi.org/10.1016/j.jbusres.2020.01.036

CIS. (2022). September barometer.

Córdoba Cubero, M. (2021). La Responsabilidad Social Empresarial y el Greenwashing como estrategias empresariales: Efectos sobre el medio ambiente. Caso de la industria textil. *Observatorio Medioambiental, 24*, 21–52.

De Jong, M. D. T., Huluba, G., & Beldad, A. D. (2020). Different shades of greenwashing: Consumers' reactions to environmental lies, half-lies, and organizations taking credit for following legal obligations. *Journal of Business and Technical Communication, 34*(1), 38–76. https://doi.org/10.1177/1050651919874105

Díaz-Soloaga, P. (2021). Communication in the fashion industry: Sustainability focus. In L. Rienda, L. Ruiz-Fernández, L. Drylie Carey, & I. García-Medina (Eds.), *Firms in the fashion industry*, (pp. 117–139). Palgrave Macmillan. https://doi.org/10.1007/978-3-030-76255-1_8

Elkington, J. (1994). Towards the sustainable corporation: Win-win-win business strategies for sustainable development. *California Management Review, 36*(2), 90–100.

Ellen MacArthur Foundation. (2021). Circular business models: Redefining growth for a thriving fashion industry. Rethinking business models for a thriving fashion industry. www.ellenmacarthurfoundation.org

European Commission. (2023). Proposal for a Directive of the European Parliament and of the Council on the substantiation and communication of explicit environmental claims (Green

Claims Directive). (22 March 2023). https://eur-lex.europa.eu/legal-content/EN/TXT/?uri=CELEX:52023PC0166

European Commission. (2021). Screening of websites for "greenwashing": Half of green claims lack evidence. (28 January 2021). www.ec.europa.eu/commission/presscorner/detail/en/ip_21_269

European Union. (2021). Climate Pact. Policy Paper. Greenwashing in the Fashion Industry. www.climate-pact.europa.eu/resources/climate-policy/greenwashing-fashion-industry_en

Fischer, D., Lüdecke, G., Godemann, J., Michelsen, G., Newig, J., Rieckmann, M., & Schulz, D. (2016). Sustainability communication. In: H. Heinrichs, P. Martens, G. Michelsen, A. Wiek, A. (Eds.), *Sustainability science* (pp. 139–148). Springer. https://doi.org/10.1007/978-94-017-7242-6_12

Genç, R. (2017). The importance of communication in sustainability & sustainable strategies. *Procedia Manufacturing, 8*, 511–516. https://doi.org/10.1016/j.promfg.2017.02.065

Gestal, I. P. (2021). Alberto Paccanelli (Euratex): La era de la deslocalización se ha terminado, Modaes.

Godemann, J. (2021). Communicating sustainability. some thoughts and recommendations for enhancing sustainability communication. In: F. Weder, L. Krainer, M. Karmasin, (Eds.), *The sustainability communication reader* (pp. 15–30). Springer VS. https://doi.org/10.1007/978-3-658-31883-3_2

Golob, U., Podnar, K., & Zabkar, V. (2023). Sustainability communication. *International Journal of Advertising, 42*(1), 42–51. https://doi.org/10.1080/02650487.2022.2144035

Guillen Mandujano, G., Vergragt, P., & Fischer, D. (2021). Communicating sustainable consumption. In: F. Weder, L. Krainer, & M. Karmasin (Eds.), *The sustainability communication reader* (pp. 263–280). Springer VS. https://doi.org/10.1007/978-3-658-31883-3_15

Han, S.LC., Henninger, C.E., Apeagyei, P., Tyler, D. (2017). Determining Effective Sustainable Fashion Communication Strategies. In: Henninger, C., Alevizou, P., Goworek, H., Ryding, D. (eds) Sustainability in Fashion, (pp. 127–149). Palgrave Macmillan, Cham. https://doi.org/10.1007/978-3-319-51253-2_7

Harlow, J., Golub, A., & Allenby, B. (2013). A review of utopian themes in sustainable development discourse. *Sustainable Development, 21*(4), 270–280. https://doi.org/10.1002/sd.522

Henninger, C. E., Brydges, T., Iran, S., & Vladimirova, K. (2021). Collaborative fashion consumption–A synthesis and future research agenda. *Journal of Cleaner Production, 319*, 128648. https://doi.org/10.1016/j.jclepro.2021.128648

INE (Instituto Nacional de Estadística). (2022). https://www.ine.es/jaxiT3/Tabla.htm?t=56934

Kantar Worldpanel. (2022). Cuestionario Lifestyles.

Kim, K. H., & Kim, E. Y. (2020). Fashion marketing trends in social media and sustainability in fashion management. *Journal of Business Research, 117*, 508–509. https://doi.org/10.1016/j.jbusres.2020.06.001

Martín López, R. (2019). Comunicación de la sostenibilidad de las principales marcas de fast fashion españolas e italianas: Zara, Mango, Calzedonia y OVS. Diferencias entre el punto de venta online y el punto de venta físico. *AdComunica, 18*, 123–154.

Martínez, P. M. (2015). Sostenibilidad, Comunicación y Valor Compartido El discurso actual del desarrollo sostenible en la empresa española. Universidad Complutense.

Milanesi, M., Kyrdoda, Y., & Runfola, A. (2022). How do you depict sustainability? An analysis of images posted on Instagram by sustainable fashion companies. *Journal of Global Fashion Marketing*. https://doi.org/10.1080/20932685.2021.1998789

Mukendi, A., Davies, I., McDonagh, P., & Glozer, S. (2020). Sustainable fashion: Current and future research directions. *European Journal of Marketing, 54*(11), 2873–2909. https://doi.org/10.1108/EJM-02-2019-0132

Newig, J., Schulz, D., Fischer, D., Hetze, K., Laws, N., Lüdecke, G., & Rieckmann, M. (2013). Communication regarding sustainability: Conceptual perspectives and exploration of societal subsystems. *Sustainability, 5*(7), 2976–2990. MDPI AG. https://doi.org/10.3390/su5072976

Pérez-Curiel, C., Jiménez-Marín, G., & García-Medina, I. (2021). The role of social media in the fashion industry: The case of eco luxury in today's consumption. In: L. Rienda, L. Ruiz-Fernández, L. Drylie Carey, & I. García-Medina (Eds.), Firms in the fashion industry (pp. 97–115). Palgrave Macmillan. https://doi.org/10.1007/978-3-030-76255-1_7

Rausch, Th. M., & Kopplin, S. F. (2021). Bridge the gap: Consumers' purchase intention and behavior regarding sustainable clothing. *Journal of Cleaner Production, 278*. https://doi.org/10.1016/j.jclepro.2020.123882

Roger, O. & Buil, P. (2014). *Manual de Comunicación Ambiental. Del greenwashing a la sostenibilidad*. Eunsa.

Salvá, A. (Ed.). (2021). Rana Plaza: los trabajadores de la industria textil de Bangladés, de nuevo en peligro. *El País*

Shen, Z. (2021). A persuasive eWOM model for increasing consumer engagement on social media: Evidence from Irish fashion micro-influencers. *Journal of Research in Interactive Marketing, 15*(2), 181–199. https://doi.org/10.1108/JRIM-10-2019-0161

Szabo, S., & Webster, J. (2021). Perceived greenwashing: The effects of green marketing on environmental and product perceptions. *Journal of Business Ethics, 171*, 719–739. https://doi.org/10.1007/s10551-020-04461-0

Vehmas, K., Raudaskoski, A., Heikkilä, P., Harlin, A., & Mensonen, A. (2018). Consumer attitudes and communication in circular fashion. *Journal of Fashion Marketing and Management, 22*(3), 286–300. https://doi.org/10.1108/JFMM-08-2017-0079

Weder, F., Karmasin, M., Krainer, L., & Voci, D. (2021). Sustainability communication as critical perspective in media and communication studies—An introduction. In F. Weder, L. Krainer, & M. Karmasin (Eds.), *The sustainability communication reader* (pp. 1–12). Springer VS.

Woodside, A. G., & Fine, M. B. (2019). Sustainable fashion themes in luxury brand storytelling: The sustainability fashion research grid. *Journal of Global Fashion Marketing, 10*(2), 111–128.

Chapter 4
Revolutionizing Textiles: Jeanologia's Tech-Driven Approach to Environmental Preservation and SDG Fulfilment

Carla Martínez-Climent, Sandra Enri-Peiró, and Sofia Arroyo

4.1 Current Situation of the Textile Industry Through the Lens of Jeanologia

Jeanologia S.L. is a large company headquartered in Valencia, Spain, with three production centres in Barcelona, Izmir (Turkey), and Vicenza (Italy). It also has subsidiaries in Hong Kong, Bangladesh, Mexico, Brazil, and the United States. This company is part of the wholesale trade industry of machinery for the textile industry and sewing and knitting machines (Axcesor, 2022). Besides technology production, Jeanologia markets software and textile consulting services. As of January 2024, the company has 201 employees from 24 different nationalities. Moreover, it has 4,400 proprietary innovations installed in 72 different countries (Jeanologia, 2023). Its international presence is strong, particularly in countries such as Pakistan, Vietnam, Japan, India, China, and Egypt, where they have local teams.

The company operates in various business lines, with the sector facing the most competition being laser technology for fabrics. In this case, Jeanologia uses laser technology to finish denim sustainably, eliminating the need for traditional practices such as scraping denim. This innovation allows giving the blue jeans the vintage denim style introduced by Levi Strauss in 1873 (Levi's, 2019). This technology can finish a pair of jeans in 90 seconds compared to the 20 minutes it takes for a manual finish. Not only that, but it also eliminates the chemicals such us the potassium

C. Martínez-Climent (✉) · S. Enri-Peiró
ESIC Business & Marketing School, Valencia, Spain
e-mail: Carla.martinezcliment@esic.edu

ESIC University, Pozuelo de Alarcon, Spain

S. Enri-Peiró
e-mail: Sandra.enri@esic.edu

S. Arroyo
Jeanologia, Valencia, Spain

J. F. Gallego-Nicholls and A. Carrilero-Castillo (eds.), *Textile Industry and the SDGs*, SDGs and Textiles, https://doi.org/10.1007/978-981-97-5073-3_4

permanganate or sodium hypochlorite to bleach denim (Alonso-García, 2022). The whole process to get the vintage look of the classic denim has been one of the most polluting practices in the entire textile industry. Therefore, Jeanologia has introduced sustainable and more efficient technologies in a market that it is in constant growth since almost 20 years. Jeanologia was one of the first companies that, in 1999, created its first laser, and a quarter of a century later, it is a leader in this sector. Interestingly, Jeanologia collaborates with major brands such as Levi's Strauss, Inditex, Tommy Hilfiger, Ralph Lauren, Jack & Jones, or Springfield, among others.

Due to the pandemic, the textile sector was severely hit by the crisis, resulting in a 30% loss in sales and a 90% loss in profit for all companies in the sector in 2020 (Salvatierra, 2021). The decline in the textile sector activity led to a 60% decrease in Jeanologia's revenue in 2021, dropping from 104.5 million euros in 2019 to 40.4 million euros in 2020, according to the consolidated accounts presented in the Mercantile Register (Torres, 2021). The subsidiary most affected by the pandemic was the company located in Italy, as the COVID-19 pandemic significantly impacted that country. This 60% decline in revenue was mitigated in 2022 when almost half of it was recovered. However, pre-pandemic levels could not be reached since many countries experienced virus rebounds. In 2023, the set goal focused on introducing ozone and color treatment technologies using a resource as simple as air (Vázquez, 2024). Additionally, the objective was to mitigate the rise in the cost of raw materials and supplies due to current wars and global inflation. Adding to these challenges, China's export closure significantly diminishes business opportunities.

It is key to keep in mind that Jeanologia sells production capacity, which we call CapEx (Capital Expenditure). Since CapEx investments in operations and production capacity are not easily affordable, Jeanologia has found a niche of customers to sell its capacity slack to (Drip Capital, 2022). Due to the substantial demand for resources, numerous customers are compelled to seek loans for financing, which poses a significant challenge owing to the high interest rates associated with borrowing. There is therefore a strong contraction of demand in the sector. At present, the company's sales have returned to growth, although without reaching pre-pandemic values. In 2024, new roadmaps have been set with a strong focus on introducing the latest solutions for dyeing and ozone washing (Silla, 2023c).

In spite of the current climate, Jeanologia is a leader in its sector thanks to its disruptive technologies and continues to work to renew itself. Its resilience and leadership in innovation has been demonstrated in the face of major economic and commercial challenges, which require effective strategies to maintain its position in the market and remain a benchmark in the industry.

4.2 Approaching the Origins of the Company

Founded in 1994 by two Valencian entrepreneurs, José Vidal and his nephew Enrique Silla (current CEO of the company), the company initially started as a textile consultancy with a team of five people (Jeanologia, n.d.). Soon, the team realized that the sustainable solutions the company wanted to provide in the textile sector required machinery to assist in the processes. Therefore, in 1999, the company launched its first laser-to-age denim, a flagship product of the company. Since then, the company has developed different processes that have helped transform the sector and make it more sustainable and ethical. In 2019, the Carlyle Group investment fund entered the company's shareholding (Pastor, 2019). One of the most renowned international funds, it acquired 40% of the company, with 35% purchased from the MCH fund and an additional 5% from its owners (Bayón, 2019). The company could continue with its projects and objectives with this capital injection.

Their mission is clear: to achieve an ethical, sustainable, and eco-efficient textile industry. Their goal is to become the preferred technological partner of their clients, providing support at every stage of their transformation and continuously adding value. Using innovative technologies, they challenge existing norms to offer their clients radical and transformative approaches. They aspire not just to improve the industry by 10% or 20%, but to achieve a significant improvement, up to 20 times better (Vázquez, 2024).

The company's vision is to become a significant agent of change in the world, positively contributing to society in all its activities (Jeanologia, 2023b). They firmly believe that companies have the power to change the world. Consequently, they are committed to actively participating in social responsibility initiatives to generate a positive impact on people and the environment.

Jeanologia started with the idea of being the best company "in the world" to becoming the best company "for the world." From laser to ozone, innovative products have been introduced that have transformed the market with the same goal: to renew the textile industry. With their current thirty years of experience, the company has evolved to encompass the entire textile finishing process with its technology.

4.3 Uncovering the Current Impact of the Company: Responsible and Sustainable Production

Jeanologia proudly asserts that it has achieved its mission by revolutionising jeans production. Currently, 50% of jeans manufactured worldwide incorporate some of the innovative technologies developed by the company (Silla, 2023a). However, they acknowledge that there are areas for improvement. The next key challenge is focused on the realm of colour and ozone. Since all textile products involve some form of colour, and pollution, water consumption, energy, and the use of toxic chemicals are intrinsically linked to this colour transfer process, Jeanologia has set a new challenge. Using advanced technology, they aim to transform how colour is applied to textiles. They have introduced a pioneering product called the "Colour Box," which allows direct colour transfer to the finished garment rather than applying it to the thread or fabric (Garcia, 2023).The Company has become one of the first to issue an ecological results report. A study that measures in cubic meters the amount of contaminated water avoided from being discharged into the rivers and seas of the planet, all thanks to their own technology. In 2023, they have saved 20.2 million m^3 of contaminated water and reduced 97,895,304 kg of CO_2 (Jeanologia, 2023c). According to the United Nations (UN), the textile sector is responsible for 20% of water pollution on the planet, and 15% of garments are never consumed (United Nations, 2019).

Previously, a hundred litres of water were required to produce a pair of jeans. Now, thanks to Jeanologia's products, this figure has been reduced to just four litres. These last four litres undergo treatment alongside the machinery before being reintegrated into the system. In this way, the company achieves what they call the "Zero Mission", an objective based on dehydrating and eliminating 100% of the waste and pollution generated in the production and finishing of jeans. This involves the elimination of water usage and detoxification, defined as the removal of toxic products in denim production, from the weaving stage to the garment's assembly, with the goal of achieving this by the year 2025 (Fig. 4.1).

Jeanologia has been in the industry for almost 30 years and recognises that they cannot achieve its goal without joining forces with the industry. All departments of the company work both internally and externally looking for allies and synergies. In 2018, Jeanologia was appointed as the official partner of denim giant Levi's Straus (Zafra, 2018). Not only does it collaborate with the brand in the United States, but its clients include companies such as Inditex, Ralph Lauren, Mango and Guess, among many others. The company actively engages and collaborates with various industry players to achieve the aforementioned goal of "Mission Zero".

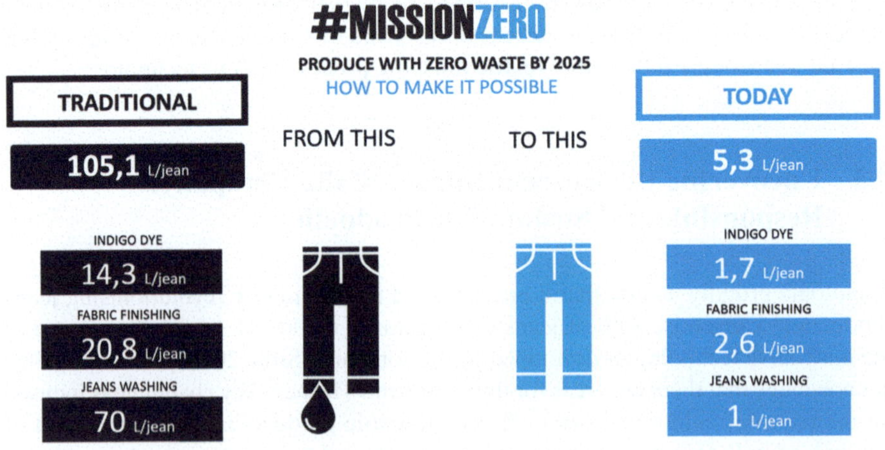

Fig. 4.1 Litres of water consumed in traditional jean finishing process vs Jeanologia process. *Source* Jeanologia (2023)

4.4 Competitive Advantage, Technology and the Role of Innovation

Jeanologia develops technologies from the beginning with the aim of eliminating traditionally polluting processes due to excessive use of water, chemicals, and energy. These processes are also unfavourable for workers. The goal is to decrease, reduce, reuse, and minimize the impact of the textile industry, which is the second-largest contributor to water pollution. The solutions presented by the company are divided into (1) technologies for fabric and/or garments, and (2) software (Jeanologia, 2023a).

- (1) Technologies

For fabric, solutions directly working on it without being sewn are called Bluescan Anubis and G2 Dynamic. The latter is a machine capable of treating fabric in the production line by applying ozone.

For garments, the product family includes laser technology to give vintage aging to denim, air washers with ozone technology (G2 ATMOS, INDRA), and water treatment with H2 ZERO. Added to the latest color treatment innovations is the Color Box technology, aimed at dyeing garments with an average saving of 60% in water and chemicals, 45% in energy, and 76% in salt during the dyeing process.

- (2) Software

The company has developed its own software for its technologies, including eMarx X for laser, an operating system called Orion, and a real-time data collection system called eData. In addition to these three products, there is the company's most recognized software, eDesigner. With this tool, laser designers can translate their creations directly onto the garment in an intuitive digital environment.

Of all the families, the laser is the most recognized product of the brand, a mature technology introduced 25 years ago, now found in factories worldwide. In the traditional denim ageing process, various manual and mechanical techniques were used, such as manual scraping, stone washing, and chemical bleaching. However, the introduction of ozone into the washing machine is noteworthy. Previously, an excessive amount of water, about 15 L per kilogram of garments, was used. Now, with Atmos technology and Jeanologia, half the amount of water and chemicals are used thanks to the nano-bubble process (Jeanologia, 2023a). The remaining water from the process is treated with H2 ZERO technology, recovering that water for 60 days.

Enrique Silla, CEO of Jeanologia, asserts that the success of their innovation process lies in "sustainably solving textile world problems." Never abandoning their goals has led them to establish an effective process that provides value to all their stakeholders (Silla, 2023a).

The process used by the company to invest and start a new project consists of approximately five phases. In each of these phases, the team must respond to a series of key questions in the following order according to Silla (2023a) (Fig. 4.2).

If any of the above questions is answered negatively, the project will not proceed. Silla claims to have received proposals from large groups and corporations, outside

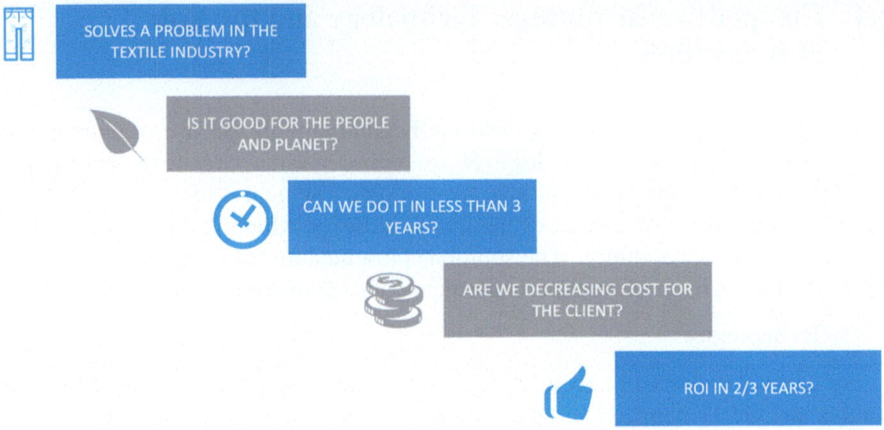

Fig. 4.2 Process of evaluation and validation of Jeanologia innovation projects (Silla, 2023a, 2023b). *Source* Jeanologia (2023)

the textile world, which it has had to reject in order to be consistent with the company's mission and vision. Since Jeanologia works with disruptive projects and cutting-edge technologies, its innovation process must follow clear guidelines. Otherwise, he says, they would be derailed as a company and leave their identity behind.

4.5 How Has the Company Oriented the Internationalisation Process?

Jeanologia is a global company operating in more than 72 countries. As mentioned earlier, the company has three production centres and has people and hub all around Asia, Africa and America.

The company is not only present in the countries where it has subsidiary companies but also collaborates with small local teams in Turkey, Sri Lanka, Vietnam, Pakistan, Egypt, Morocco, South Africa, among others (Segura, 2023).

Jeanologia operates as a Business to Business (B2B) company, and its clients include laundries and textile finishing factories working with major corporations worldwide. Each mentioned production centre specializes in a different product. Laser production takes place in Sant Quirze del Vallés (Barcelona) due to the city's historical expertise in this field. In Italy, there is production for washers, and in Turkey, the focus is on technology related to ozone.

The company underwent a rapid internationalization process because of the need to be close to customers' required local presence. Jeanologia emphasizes local employability and talent in all its locations. It has multidisciplinary and international teams working towards a common and clear goal of sustainability (Segura, 2023).

In 2021, the organization decided to embark on a new project called "Urban Factories," an initiative aiming to reintroduce production in Europe through the implementation of technology and automated processes (Vázquez, 2024). Since most textile garments are manufactured in Asia, where workers earn less than 200 euros per month, the strategy is to automate the process and simultaneously ensure sustainability by moving production to Europe. However, Enrique Silla admits that the Eurozone is not prepared for the change, and the concept of "Urban Factory" may face challenges in penetrating the market. Nevertheless, it is a project already underway in the United States in collaboration with Levi's, which is using this "factory" as a testing centre.

4.6 Company Contribution to the Sustainable Development Goals (SDGs)

Jeanologia attributes its success to having a clearly defined mission and purpose, such as "People and the planet before profits," reflecting the essence of its philosophy where priority is given to the well-being of the planet and people over economic gains. The company's strategy is built on three fundamental pillars: People, Planet, Product, addressing essential aspects such as eco-efficiency, digitization, sustainability, and automation. This is achieved with the help of a well-prepared and dedicated human team that combines a passion for fashion with a solid technological foundation, integrating creativity and science. Their achievement involves overcoming barriers between the artistic and the scientific, inventive and technological, by combining imaginative and scientific minds. This approach entails an appreciation for the product, fashion, and creativity, coupled with the application of technology, rigor, and a scientific approach, all directed toward a common goal.

Since 2019, Jeanologia has participated in the Global Compact regarding the Sustainable Development Goals (SDGs) and provides an annual report on its progress in this field (United Nations Global Compact, 2023). It is crucial to note that the organization has a significant impact on various SDGs. Specifically, SDGs 6 (Clean water and sanitation), 9 (Industry, innovation, and infrastructure), 12 (Responsible consumption and production), 13 (Climate action), and 17 (Partnerships for the goals) reflect its commitment to sustainability and corporate responsibility.

According to the latest SDG progress report in 2022, the company has supported all the goals of the 2030 agenda, each with at least one action. The following are the most significant contributions based on Jeanologia (2023).

SDGs 1 (No poverty) and **2** (Zero hunger): The company collaborates with eight non-profit organizations, including the project in Burkina Faso with the Friends of Rimkieta Foundation. Through this collaboration, they have funded projects, including the construction of three wells, a kindergarten, and a garden aimed at the job integration of women. In Rimkieta, the wells provide access to clean water and irrigation for the fields, crucial for the region's development. Additionally, they collaborate with NGOs such as Asindown or Extraordinary Projects in Valencia, Spain. A total of over one hundred thousand euros invested in various charitable causes (Jeanologia, 2023, 2). Not only do they participate, but they also encourage their employees, known as Jeanologists, to participate in these projects. The company offers an extra week of vacation to employees to invest in charitable actions, a program they call "1 + 1."

SDGs 6 (Clean water and sanitation) and **14** (Life below water): In 2023, thanks to the company's technologies, 20.2 million cubic meters of contaminated water have been saved. The ozone technology reduces water usage from 100 to 5 L, and these remaining 5 L, treated by the H2 Zero technology that reuses them, are incorporated back into the process. They not only support this goal with their technologies but also with the wells built in Burkina Faso, which irrigate the fields in the area and provide access to clean water for many families.

SDG 12 (Responsible consumption and production): Through their major clients in the industry such as Levi's, Mango, or Jack&Jones, among others, Jeanologia enables them to market more sustainable products with a lower environmental impact. Relying on sustainable technologies and Jeanologia's EIM software, which allows brands to label their garments based on their energy, water, chemical usage, and worker health. With this information, clients have access to accurate and trustworthy information.

SDGs 13 (Climate action) and **17** (Partnerships for the goals): The company is dedicated to raising awareness among clients, employees, and society in general about the importance of applying environmentally friendly technologies that contribute to caring for the planet. This is achieved by reducing water consumption and the use of chemicals in textile finishing processes. The company considers environmental protection and climate fundamental, which is why the dissemination and awareness of their vision are key aspects. In this regard, various actions have been taken in collaboration with the CRE100DO association, the MISSION ZERO chair at the University of Valencia, and other centres both nationally and internationally. In addition to these actions, participation in international fairs specialized in the textile sector contributes to spreading a sustainable alternative to the market.

All the information described above is published on the official website of the Spanish Global Compact (2023). Their Sustainability Report is also available and published on their website (Jeanologia, 2020). All of this demonstrates a transparent and proactive approach to disclosing their sustainable practices. The inclusion of ESG metrics and the use of tools like Microsoft Power BI to collect and present data highlight the importance Jeanologia places on measuring and continuously improving its environmental, social, and governance performance. Additionally, the formation of the Sustainability Committee in 2019, composed of personnel from various departments (Marketing, Innovation, and Internationalization), underscores the multidisciplinary and internal collaboration to address sustainability issues throughout the organization (Segura, 2023).

One of the most disruptive solutions introduced by Jeanologia in recent years is called EIM (Environmental Impact Measurement). It is defined as measurement software that allows monitoring the environmental impact of garment finishing processes

(Fariña, 2023). EIM can measure water and energy consumption, chemical impact, and worker well-being. It scores each process based on an average of the four categories, with all of them carrying the same weight. With this product, Jeanologia helps its clients be aware of their impact and have the ability to inform their customers and stakeholders about their results. Internationally recognized brands such as H&M, Marks & Spencer, or Jack & Jones are already users of EIM (C. Silla, personal communication, December 15, 2023).

4.7 Conclusions and Contributions

Jeanologia is a company that, even with 30 years of experience, maintains its startup spirit with a team eager to continue revolutionizing the denim industry. Its success lies in many factors that, together, make it a conscious company, with a product designed from the beginning to be sustainable and take care of the planet and people. Jeanologia remains a Valencian family-owned company that has never stopped looking beyond its borders, always travelling with its mission.

The Sustainable Development Goals of the 2030 Agenda have been created as a guide for companies, NGOs, governments, and individuals. Jeanologia incorporates these goals not only into its mission, vision, and culture but also into its production process and as a value proposition for its clients. Jeanologia is an agent of change. Its success is also based on establishing a creation process focused on the textile industry and having a team passionate about caring for the planet.

For the textile industry to align with the objectives of the 2030 Agenda, it is vital to promote projects like Jeanologia. A medium-sized company that can have a very positive impact on the industry. Not all companies can replicate an identical business and production model. However, it does not mean they cannot contribute to fulfilling the SDGs by including them in their missions and visions and communicating their positive impact to society.

Through this case study, we seek to give visibility to good practices associated with the textile industry from a practical approach. In this way, we seek to encourage the scalability of this type of project, thus generating a gradual growth of business projects with sustainable impact.

References

Alonso-García, E. (2022). *Efectos de envejecimiento en tejido denim mediante tratamientos enzimáticos*. In UPV. Escuela Politécnica de Valencia. https://riunet.upv.es/bitstream/handle/10251/186568/Alonso%20-%20EFECTOS%20DE%20ENVEJECIMIENTO%20EN%20TEJIDO%20DENIM%20MEDIANTE%20TRATAMIENTOS%20ENZIMATICOS.pdf?sequence=1

Axcesor. (2022). *Jeanologia S. L-Telf 963891092*. https://www.axesor.es. https://www.axesor.es/Informes-Empresas/798589/JEANOLOGIA_SOCIEDAD_LIMITADA.html

Bayón, Á. (2019). *El gigante del capital riesgo Carlyle entra en la empresa española que "desgasta" los vaqueros.* Cinco Días. https://cincodias.elpais.com/cincodias/2019/03/12/companias/1552416459_887139.html
Drip Capital. (2022). *¿Qué es Capex?|Significado, Ejemplos y Cómo se Calcula.* https://www.dripcapital.com. https://www.dripcapital.com/es-mx/recursos/finanzas-guias/que-es-capex
Fariña, N. (2023). *La moda planta cara al lado más oscuro del color.* El País. https://elpais.com/sociedad/moda-futuro-y-accion/2023-11-13/la-moda-planta-cara-al-lado-mas-oscuro-del-color.html
Garcia, B. (2023). *Entrevista con project manager de Jeanologia* (S. Arroyo, entrevistadora) [Comunicación verbal].
Jeanologia. (2020). *Innovaciones textiles sostenibles para la industria-Jeanologia.*
Jeanologia. (2022). Sostenibilidad. https://www.jeanologia.com/es/sustainability/
Jeanologia. (2023a). *Cuenta de resultados sostenible de ahorro en agua y CO2.*
Jeanologia. (2023b). *Jeanologia catálogo de tecnologías.* Jeanologia. https://www.jeanologia.com/
Jeanologia. (2023c). *Sobre nosotros.* Jeanologia.com. https://www.jeanologia.com/es/nosotros/
Jeanologia S.L. (n.d.). *Empresa de soluciones textiles sostenibles- Sobre Jeanologia.* Jeanologia. Retrieved January 16, 2024, from https://www.jeanologia.com/es/nosotros/
Jeanologia S.L. (2023). Informe Progreso ODS Jeanologia S.L. In *Global Compact United Nations.* Pacto Mundial de Naciones Unidas. https://unglobalcompact.org/what-is-gc/participants/137504-JEANOLOGIA-SL
Levi's. (2019). *Levi's® ES.* Levi.com. https://www.levi.com/ES/es_ES/features/about-us
Pastor, E. (2019). *Carlyle Group entra en el accionariado de la compañía valenciana Jeanología* (EDICIONES PLAZA, Ed.). Valencia Plaza. https://valenciaplaza.com/carlyle-group-entra-en-el-accionariado-de-la-compania-valenciana-jeanologia
Salvatierra, J. (2021). *Crisis en la industria textil: Un 2021 con la moda de hace un año.* EL PAÍS. https://elpais.com/economia/2021-02-28/crisis-en-la-industria-textil-un-2021-con-la-moda-de-hace-un-ano.html
Segura, E. (2023a). *Entrevista con responsable del departamento internacional de Jeanologia* (S. Arroyo, entrevistadora) [Comunicación verbal].
Silla, C. (2023a). *Entrevista con responsable del marketing de Jeanologia* (S. Arroyo, entrevistadora) [Comunicación verbal].
Silla, E. (2023b). *Presentación sobre Jeanologia a la asociación de empresarios valencianos.* Visita empresarios valencianos.
Torres, B. (Ed.). (2021). *Jeanologia factura un 60% menos y reduce su beneficio un 85% por el cierre mundial del comercio.* Valencia Plaza. https://valenciaplaza.com/jeanologia-reduce-un-60-su-facturacion-por-el-cierre-mundial-del-comercio-en-la-pandemia
United Nations. (2019). *La Alianza de la ONU para la Moda Sostenible abordará el impacto de la "moda rápida."* UN Environment. https://www.unep.org/es/noticias-y-reportajes/comunicado-de-prensa/la-alianza-de-la-onu-para-la-moda-sostenible-abordara-el
Vázquez, J. (2024). *"Para traer la producción a Europa necesitamos automatizarla y hacerla sostenible."* Levante-EMV. https://www.levante-emv.com/economia/2024/01/06/traer-produccion-europa-necesitamos-automatizarla-96573253.html
Zafra, I. (2018). Así es la tecnología española que hará sostenible a Levi's. *El País.* https://elpais.com/economia/2018/03/02/actualidad/1520016938_986293.html

Chapter 5
From Shoppers to Agents of Change: Consumer Behavior and Collaborative Fashion Consumption in Spain

Silvia Blas Riesgo

5.1 Introduction

Although worries about resource scarcity and the underlying relationship between humankind and Earth have accompanied society since its early days, sustainability has only become a dominant narrative in recent times. Globally, a sustainability plan to be achieved by 2030 called the Agenda 2030 has been drafted. Within the Agenda 2030 (United Nations Sustainable Development Goals, 2015), this research focuses on the Sustainable Development Goal number 12 (SDG12): "Ensure sustainable consumption and production patterns". SDG12 considers that sustainable consumption and production patterns are vital mechanisms for improving living standards without compromising future generations' resource needs.

Decoupling economic growth from resource use is one of the most complex challenges facing the world today since we continue to use ever-increasing amounts of natural resources to support our economic activities globally (United Nations, 2019). Hence, doing so effectively will require a profound transformation of business practices along global value chains and consumer lifestyles and behavior.

From the given definition of SDG12 by the UN, we can individualize three kinds of actors from which action is required: governments, businesses, and individuals. While governments and businesses must provide adequate cultural and market frameworks, aligning individuals' consumption behavior to more sustainable patterns is essential to achieving global sustainability goals.

This research mainly focuses on consumer behavior applied to the field of fashion. The fashion industry can be considered responsible for reinforcing some of the most damaging consumption practices of consumers by promoting, through its business

S. Blas Riesgo (✉)
Department of Business Administration, Marketing in the Digital Economy, University of Zurich, Zurich, Switzerland
e-mail: silvia.blasriesgo@business.uzh.ch

J. F. Gallego-Nicholls and A. Carrilero-Castillo (eds.), *Textile Industry and the SDGs*, SDGs and Textiles, https://doi.org/10.1007/978-981-97-5073-3_5

models, practices of overconsumption and waste. The 'throw-away' paradigm, exacerbated by fast fashion and its latest evolutions–ultra-fast fashion and low cost–has led to negative environmental, societal, economic, and cultural consequences.

Amid the surging wave of sustainable fashion, it is imperative for businesses to grasp the motivating factors behind consumer choices. Consumers' engagement in sustainable consumption practices is essential to achieving general sustainability goals; with an accurate understanding of consumers' purchasing patterns, industry actions toward sustainability could succeed. However, to this date, consumer attitudes, drivers, and barriers regarding sustainable fashion consumption remain relatively under-researched, especially in the Spanish context.

For this study, sustainable fashion consumption is considered not only purchasing brand-new sustainable fashion products but also participating in new business models that embrace circularity, such as Collaborative Fashion Consumption (CFC)–within which secondhand, renting or leasing models–and embracing the Slow Fashion philosophy.

Against this backdrop, this research explores how Collaborative Fashion Consumption can be a way to bring the fashion industry closer to the achievement of SDG12. To do so, we explore fashion consumers' behavior towards secondhand and renting business models in Spain through a series of focus groups (specifically, three focus groups with 23 participants).

The rest of the chapter is divided as follows. First, we present a literature review defining sustainable fashion consumption and what it entails, why the current consumption model in the fashion industry is unsustainable, and possible alternative business models. Afterward, we present the methodology and main findings to culminate with the discussion, implications and conclusions and future research agenda.

5.2 Literature Review

5.2.1 *What is Sustainable Fashion, and What does Sustainable Fashion Consumption entail?*

Sustainable products have been defined in the literature as products that provide environmental, societal, and economic benefits while protecting public health, welfare, and the environment over their entire commercial cycle - from the extraction of raw materials to final disposal - providing for the needs of future generations (Pencarelli et al., 2020). Thus, Mohr et al. (2001) defined sustainable consumption as the pattern of purchasing and consuming products that maximize long-term benefits and minimize hazardous effects on consumers and societies.

For this research, we consider sustainable fashion as a system–*The Sustainable Fashion System* (Fig. 5.1)–to encompass the holistic nature of sustainable fashion and highlight the role of each of the system's actors: governments, businesses and consumers. In social sciences, the *International Encyclopedia of the Social & Behavioral Sciences* (2001) defines *a system* as "The interconnected relationship between individuals, groups, and institutions with shared behaviors, norms, and values that combine to form society". Thus, sustainable fashion is only possible if governments, fashion companies and consumers *work together* in interdependence toward a marriage between ethics and aesthetics.

Donella H. Meadows (in Fletcher, 2008) cites an ancient Sufi tale to explain the principle of interdependence better. A story from the *Tales of the Dervishes* talks about people who could not see a whole elephant, so they made assumptions about it from the separate parts they could touch. As a result, those who felt its ear claimed that the elephant was "wide and broad like a rug," others who touched the trunk believed it was "like a straight and hollow pipe", and yet others who touched the elephant's leg said that it was "mighty and firm, like a pillar". The moral of the tale is that focusing only on one part, in isolation from others, does not lead to an accurate understanding of the whole. Accordingly, each part has its place among others, and together, they contribute to the functioning of the whole.

Therefore, companies and consumers both hold responsibilities, legal and moral, under the policy and cultural framework established by governmental institutions–as the three leading actors in sustainable production and consumption appointed by SDG number 12- and these three leading actors need to coordinate efforts:

- Governments should promote policies that stimulate companies to behave sustainably (e.g., by applying tax incentives to reward companies that behave sustainably).
- Companies must work on the social and environmental side to deliver the most sustainable products while retaining financial gains (e.g., fashion companies could develop their own platforms for repurposing and monetizing pre-owned fashion items).
- Consumers must be conscious about their purchases and elongating the lifecycle of those products: what they buy, how often, how they use those products, and how they dispose of them (e.g., consumers could try to control impulse buying by assigning a monthly budget to fashion consumption, which could lead not only to more sustainable habits but also to economic savings and avoiding post-shopping regrets).

5.2.2 *The Issue of Overconsumption and Waste in the Fashion Industry*

The fashion industry can be considered responsible for reinforcing some of the most damaging consumption practices of consumers by promoting, through its business models, practices of overconsumption and waste. The 'throw-away' paradigm, exacerbated by fast fashion and its latest evolutions–ultra-fast fashion and low cost–has led to negative environmental, societal, economic, and cultural consequences.

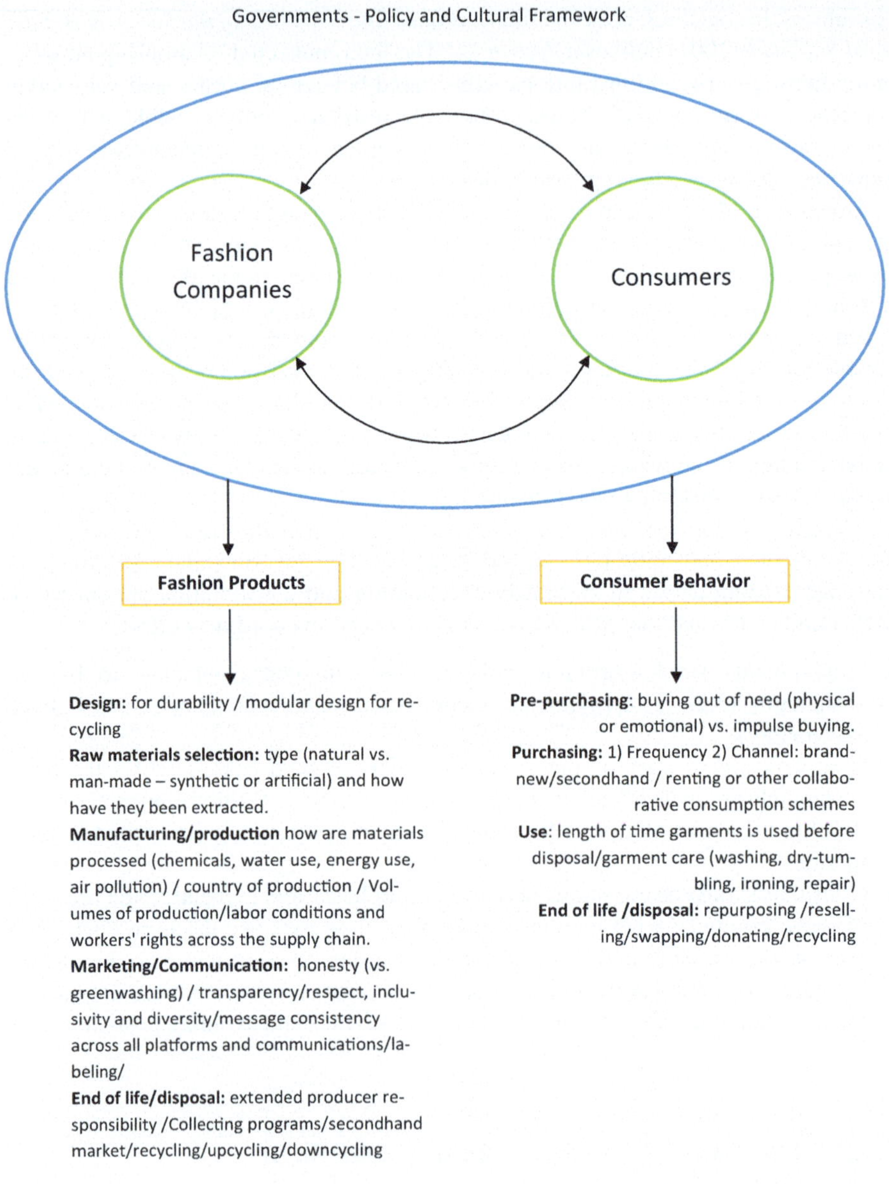

Fig. 5.1 The sustainable fashion system

The mass production of fashion garments has existed since the mid-nineteen century; however, the advent of fast fashion business models since the 1970s, the liberalization and globalization of markets, increased digitalization, and the preponderance of image-based social media have disrupted the pace of both fashion production and consumption (Niinimäki et al., 2020). Fast fashion and, to a greater extent, its latest evolutions–ultra-fast fashion and low cost–rely on mass production, cheap manufacturing, low prices, frequent consumption, large sales volumes, and short-lived garment use (Niinimäki et al., 2020; Fletcher, 2016; Caro & Martínez-de-Albéniz, 2015).

In recent years, ultra-fast-fashion brands have gone from virtually unknown to rivaling the incumbent category's most prominent players. For instance, according to The Business of Fashion (2022a, 2022b), *Shein* is reportedly raising funding at a valuation of 100$ billion, which would make *Shein* worth as much as *Zara* and *H&M* combined. One of the primary reasons for *Shein's* rapid growth is that it launches more than ten times the number of styles than all its main competitors combined. According to The Business of Fashion (2022a, 2022b), until April 2022, *H&M* launched 4,414 new styles, *Zara* 6,849, *Boohoo* 18,343 and *Shein* 314,877.

The following paragraphs briefly highlight the impact of overconsumption and waste on the economy, the environment, culture, and society.

Many factors have contributed to the surge in fashion consumption over the past few decades, but a prominent one is the substantial decline in garment prices. Technological advancements, increased production efficiency, and globalized supply chains have collectively driven down the cost of clothing to unprecedented levels. Between 1996 and 2018, clothing prices in the European Union (EU) fell by over 30% relative to inflation, which has fueled increased consumption and reduced clothing lifespans (European Environment Agency, 2019).

Paradoxically, despite owning a growing number of garments, the EU's average per-person expenditure on clothing and footwear has decreased significantly. From 30% in the 1950s, it dropped to 12% in 2009 and further to 5% in 2020 (Jackson & Shaw, 2008; Sajn, 2019). This trend is mirrored globally, with Peters et al. (2019) estimating that global per-capita textile production has nearly doubled, from 5.9 to 13 kg per year, between 1975 and 2018.

The fashion industry's relentless pursuit of lower prices and faster production cycles has had far-reaching environmental and social consequences. The industry's massive production volumes contribute significantly to greenhouse gas emissions, water consumption, hazardous chemical releases, and biodiversity loss. In 2017, the production and handling of clothing and footwear purchased by EU-28 households consumed an estimated 1.3 tons of primary raw materials and 104 cubic meters of water per person (European Environment Agency, 2019).

The trend towards lower production costs has also led to a geographical shift of garment manufacturing to developing countries, primarily in East Asia. This shift is driven by the desire to capitalize on lower labor costs and lax environmental regulations in these regions. As a result, 60% of garment production is concentrated in East Asia, while about 60% of consumption occurs in the EU, US, Japan, and China. Unfortunately, this shift has often come at the expense of workers' rights and safety. The Rana Plaza factory collapse in Bangladesh in 2013, which claimed the lives of over 1,100 garment workers, tragically highlighted the dire working conditions that persist in many developing countries.

Wages in garment production countries often fall below the minimum wage, failing to provide workers with a living wage. While the garment and textile sector may offer higher wages than traditional sectors like agriculture, over half of the workers in the fashion industry are not paid a minimum wage. Moreover, even in countries with minimum wage laws, these wages often fall significantly short of what is considered a living wage (Global Living Wage Coalition, 2021; ILO, 2015).

In conclusion, the declining cost of fashion has undoubtedly fueled increased consumption and contributed to the industry's global dominance. However, this has come at a substantial environmental and social cost. The industry's unsustainable practices have resulted in widespread pollution, resource depletion, and worker exploitation. Addressing these issues will require a fundamental shift in the fashion industry's approach to production, consumption, and labor practices.

5.2.3 Alternative Sustainable Fashion Business Models

This section highlights the main alternative business models that could help the fashion industry achieve its sustainability goals. As has been explained, the rise of fast fashion as a business model–by relying on low prices and ever-increasing sales volumes–has significantly enlarged the detrimental impacts of the fashion industry towards sustainability. On the other hand, the increasing awareness of the importance of sustainability within the fashion industry has led to new business practices and, therefore, new ways of consumption.

Sustainable fashion consumption exceeds the consumption of brand-new sustainable fashion products, and thus, it entails how consumers use, care, and dispose of fashion garments. Here, I present an overview of the alternative sustainable business models in the fashion sector–the Circular Economy, the Sharing Economy or Collaborative Fashion Consumption and Slow Fashion–and its capacity to help move the fashion industry towards sustainability by contributing to achieving sufficiency and efficiency goals.

How businesses operate strongly influences how products are produced and consumed (Bocken et al., 2014; Boons & Ludeke-Freund, 2013). While producing more brand-new sustainable clothing has been a prevalent strategy in the sustainable fashion movement, its effectiveness may be limited, given the sheer volume of garments being purchased and discarded (Becker-Leifhold & Iran, 2018). As a result, companies in the fashion industry are increasingly turning to innovative, sustainable business models to address the challenges posed by excessive clothing consumption and disposal (Kant Hvass, 2015).

The purpose of the circular economy is to elongate products' life span, keeping products and materials at their highest utility and value for the longest possible time to avoid constantly producing new products. To do so, there are two main pathways: recycling the materials/products through chemical and engineering processes–which appertains more to the industrial domain–or reusing products by reincorporating them in the market in their current state once the owner is ready to discard them, which depends more on consumers' behavior–see Fig. 5.2.

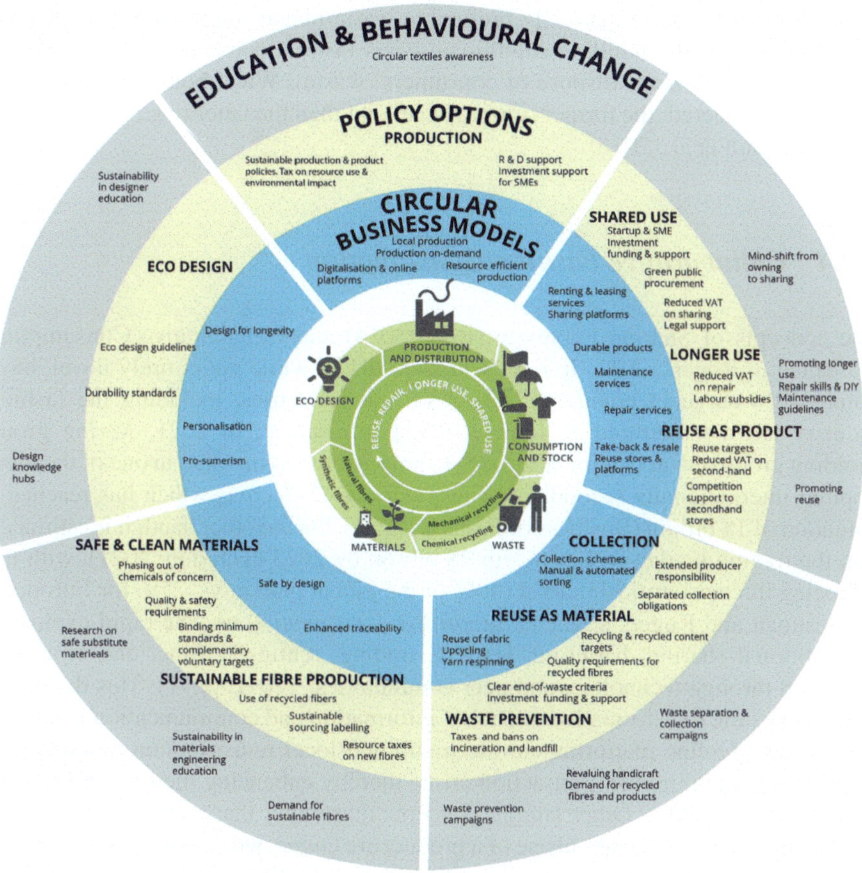

Fig. 5.2 The role of circular business models, policy options and behavioral change in circular textiles systems. *Source* European Environment Agency (2019)

Meanwhile, the reuse category has also been termed the Sharing Economy (Belk, 2014) or Collaborative Fashion Consumption (Botsman & Rogers, 2011; Iran & Schrader, 2017), which entails that the ownership of a fashion garment passes from one consumer to another (secondhand, sharing, swapping or lending), or even no one has the ownership of the garment, but several consumers have use-access (through renting and leasing schemes). Companies willing to engage in the Sharing Economy can provide a platform for consumers to exchange their garments (the company provides the platform, and the exchange is peer-to-peer, like *Wallapop*) or can directly resell, rent, or lease the garments themselves (Business to Consumer, or B2C).

Regarding the benefits of these innovative business models towards achieving sustainability, Sandin and Peters (2018) conducted a literature review on the environmental impact of textile reuse and recycling. After analyzing 41 research items, the authors conclude that the literature provides strong support for the claim that textile reuse and recycling are generally preferable waste management options compared to incineration and landfilling (traditional ways fashion companies discard unsold goods and governments dispose of consumers' waste). When both reuse and recycling are considered, the former is more beneficial than the latter (e.g., Dahlbo et al., 2017; Schmidt et al., 2016).

5.2.4 Collaborative Fashion Consumption

The concepts of Sharing Economy (Belk, 2014) and Collaborative Consumption (Botsman & Rogers, 2011; Iran & Schrader, 2017), while not entirely novel, have witnessed a surge in popularity across various industries, including the fashion industry (Becker-Leifhold & Iran, 2018). For instance, in 2021, *Kering* group (holding group of *Gucci, Balenciaga, YSL,* among others) invested in one of the most popular internationally secondhand sites, *Vestiare Collective*, which has reached a valuation of 1$ billion (Business of Fashion, 2021). On the other hand, it is estimated that the renting market will be worth 2$ billion by 2023 (Thredup, 2021), with the US being the most critical market (40% of the global market vs. 27% for Europe).

Bostman and Rogers define *collaborative consumption* as "the rapid explosion in traditional sharing, bartering, lending, trading, renting, gifting, and swapping redefined through technology and peer communities" (2011, p. 71). This definition underscores the transformative impact of information and communication technologies (ICTs). Online platforms elevate traditional local redistribution markets to a wider reach and diminish transaction costs, thereby enhancing their value proposition (Schrader, 2001). Other similar concepts that imply that consumers renounce ownership and share the use phase of a product are connected consumption (Schor & Fitzmaurice, 2015) and access-based consumption (Bardhi & Eckhardt, 2012).

Applied to the fashion sector, Iran et al. (2017) define *Collaborative Fashion Consumption* as a consumption trend "in which consumers, instead of buying new fashion products, have access to already existing garments either through alternative opportunities to acquire individual ownership (gifting, swapping, or second hand) or through usage options for fashion products owned by others (sharing, lending, renting, or leasing)". The different forms of collaborative fashion consumption can be broadly categorized into two types: peer-to-peer (P2P) and business-to-consumer (B2C) (Iran & Schrader, 2017).

While alternative approaches to acquiring clothing have existed for some time, they have reached unprecedented levels of adoption and expansion due to the advancements in information and communication technologies (ICTs). Historically, CFC had been largely confined to local exchanges within families, social circles, or between acquaintances. For instance, Marshik (2011) points out that at the beginning of the twentieth century, secondhand consumption was a common practice among fashion aficionados, although it was performed with discretion because it was a stigmatized practice.[1] Today, various online and offline platforms enable, expedite, and broaden the reach of CFC among a wider consumer base. Consequently, CFC is no longer solely confined to peer-to-peer (P2P) exchanges but can be fostered and orchestrated by businesses as well.

Collaborative fashion consumption emerges as a promising avenue for the fashion industry to reduce its environmental footprint by closing material loops, minimizing resource consumption, curbing waste generation, and elongating the product use phase (life span). Advocates of Collaborative Fashion Consumption highlight that from a sustainability perspective, reusing garments significantly diminishes the environmental impact of a garment's life cycle (Dahlbo et al., 2017; Sandin & Peters, 2018). This is attributed to the substantial energy savings associated with reusing garments compared to the production of new ones (Bras-Klapwijk et al., 2001; Fletcher, 2008).

According to Waight (2013) and Woolridge et al. (2006), reuse emerges as a compelling strategy for promoting a greener economy by significantly reducing the environmental footprint of clothing consumption compared to purchasing newly manufactured apparel from virgin resources. Bostman and Rogers (2011) underscore that even when accounting for transportation emissions and resource utilization associated with the reuse model, it still holds a substantial environmental advantage over producing a new product. This suggests that reuse offers a viable and sustainable avenue for reducing the environmental impact of the fashion industry.

5.2.5 *The State of Collaborative Fashion Consumption in Spain*

According to a report by the secondhand outlet Micolet (2020), 28% of Spanish women buy secondhand garments. The report claims that the main motivations to buy secondhand are: (1) Saving money (economic reasons), (2) Access to exclusive items otherwise unattainable, and (3) Sustainability motivations: 15% of women claim to consider environmental factors in the fashion purchase decision making. The profile of secondhand users is: 70% of the Generation Z assured to have bought and/or sold secondhand items, 36% among Millennials, and 46% among Baby Boomers; from 2019 to 2020, Millennials and Baby Boomers have been the segments which have

[1] Marshik outlines that secondhand clothing exchanges were shared among all social classes at the beginning of the twentieth century. However, it had to be performed with discretion since this practice already had a stigma. For instance, upper- and middle-class women managed secondhand sales through a middlewoman. Marshik remarks that while not all women of the upper class were financially affluent, maintaining a respectable appearance was still a significant social expectation. To acquire new garments, many women relied on generating income from the sale of their used clothing—a practice that indirectly benefited those with less means. Marshik explains: "Many women only came into contact with clothing designed by Chanel and Poiret after other women had worn the garments and decided to sell them".

increased more their secondhand acquisitions. The top ten brands sold most often via *Micolet* are (1) *Zara*, (2) *Stradivarius*, (3) *Bershka*, (4) *Mango*, (5) *H&M*, (6) *Pull&Bear*, (7) *Primark*, (8) *Lefties*, (9) *Sfera*, and (10) *Massimo Dutti*.

Currently, there are a variety of companies selling secondhand fashion in Spain:

- Business to Consumer (B2C): In physical stores, several NGOs such as *Humana*, *Cáritas' Moda Re'* or *Intermón Oxfam*, and some little-known neighborhood stores, with higher concentration in the big cities such as Madrid (like *Milmodelitos*), Barcelona or Valencia. Online: *Micolet*, *Percentile*, *Vinted* (*Vinted* is a Berlin-born company that has entered the Spanish market by acquiring the secondhand online outlet from Málaga *Chicfy*), Adoptaunaprenda, and *SecondBaby* (for infants' garments).
- Consumer to Consumer (C2C): There are no platforms solely specializing in fashion, but they sell all kinds of products. *Wallapop* is the most popular, *followed by Vinted, Vibbo* (previously known as *Segundamano*), and *Milanuncios*.

Renting and subscription models have also been flourishing in Spain in recent years. These business models are, for now, mainly and exclusively directed to women. Some notable companies offering these services (B2C) are *Ecoadicta*, *Pantala*, *La Más Mona*, *Armario Infinito*, *Pislow*, *¿Me lo prestas?*, *Lamodateca*, *Ouh Lo Là* or *Verone* (see Canalejo, 2020). A slightly different model is *Modateca*, where you can donate the clothes, you do not use anymore and receive credit in exchange for renting the garments they offer. Furthermore, there is also the possibility of renting C2C; for instance, in *Jointly*, you can rent from others or let others rent your clothes for four days.

Unfortunately, no data has been found regarding the revenues or the share of the secondhand and renting and subscription business models in Spain to understand to what extent consumers are engaged with these fashion consumption alternatives. Some indicative figures are that, for example, *Humana* in 2015 had 40 stores distributed throughout the Spanish territory, which sold 2.8 million secondhand fashion garments; in 2019, they had 53 stores and sold 5.2 million garments (Retema, 2021). Thus, *Humana* claims that sales have increased by 38% in their secondhand outlets, and the number of consumers has increased by 20%. Regarding renting, *La Más Mona* (founded in 2012) had revenues of 620,000€ in 2019 (modaes.es, 2020), and *Ouh-Lo-Là* claimed to have 3,000 subscribers up to 2020 (Reyes et al., 2020).

5.3 Methodology

Three focus groups were conducted in Madrid, Spain, to explore consumer perceptions and motivations regarding Collaborative Fashion Consumption (CFC) models, particularly renting and secondhand business models. Participants were recruited through convenience sampling, focusing on individuals interested in fashion.

The first and second focus groups (n = 8 and n = 9, respectively) consisted of fashion consumers with varying levels of fashion interest but no prior experience with sustainable fashion. These groups were further divided by age: one subgroup included participants under 30 years old, while the other included participants 30 years of age and older.

The third focus group (n = 6) comprised individuals actively engaged with sustainable fashion practices. These participants, ranging from 21 to 38 years old, were recruited through collaborations with Fashion Revolution Spain, Slow Fashion Next, Humana NGO Spain, and the Spanish Association of Sustainable Fashion (AMSE).

The focus group sessions aimed to achieve the following objectives: (1) Reveal consumption habits and shopping preferences in fashion, (2) Define the frequency and preferred channels to buy fashion, (3) Understand participants' knowledge of sustainable fashion, (4) Identify motivations and barriers to engaging in CFC models, particularly renting, leasing, and secondhand business models, (5) Gather insights into consumer perceptions and preferences regarding CFC practices.

Each focus group session lasted approximately 90 min and followed a structured format, gradually deepening into the topic of CFC. To avoid priming effects, words such as "sustainable," "ethical," "eco," or "green" were deliberately avoided in the initial stages of the discussion.

Data collection was primarily focused on the individual and group levels of analysis, as outlined by Cyr (2019). The individual level allowed for capturing multiple individual opinions simultaneously, while the group level facilitated the identification of consensus on key topics. Additionally, focus group interactions were carefully observed to uncover unexpected ideas or insights that could inform future research directions.

Thus, the data gathered from the focus groups were transcribed verbatim and analyzed using NVivo 12 software. An inductive methodology was utilized, adhering to the general procedural model introduced by Mayring (2010). The content analysis involved four main steps: (1) Transcription of focus group discussions in verbatim, (2) Incorporation of researcher notes and external cues into the transcripts, (3) Inductive identification of categories and themes through thoroughly examining the text, and (4) Development of a structured coding scheme to organize and classify the identified themes, categories, subcategories, and codes.

This systematic approach to data analysis ensured the extraction of meaningful insights from the focus group discussions, providing valuable information about consumer perceptions and motivations surrounding CFC models.

5.4 Results

5.4.1 Overview

Participants in FG1 (<30) were the most hesitant to embrace the secondhand market, primarily citing a lack of trust, which will be further discussed below. They had no prior experience with secondhand purchases and had only sold items occasionally through the peer-to-peer platform *Wallapop.* Conversely, participants in FG3 strongly preferred the secondhand market, actively engaging in both buying and selling. They prioritized the secondhand market over purchasing brand-new sustainable fashion items, emphasizing the importance of product circularity, and extending the lifespan of garments as the most sustainable approach.

Half of the participants in FG2 (>30) reported engaging in secondhand purchases, albeit at varying frequencies. Some participants were frequent shoppers, while others only purchased secondhand items occasionally. Economic motivations were the primary driving force behind their secondhand purchases, particularly during periods of personal financial hardship. However, some participants were also motivated by sustainability concerns, seeking to minimize the production of new garments and contribute to extending the lifespan of clothing, thereby reducing waste and environmental impact. Notably, none of the focus group participants had experience renting garments or accessories.

Table 5.1 below summarizes the main differences among focus group participants in their sustainability approach and fashion shopping decision-making process.

5.4.2 Barriers and Drivers to Participating in Collaborative Fashion Consumption

This section will first present the barriers to engaging in Collaborative Fashion Consumption identified during the focus group sessions and provide an overview of the identified barriers by classifying them into individual traits, product attributes, and situational variables, and later will do the same with the main drivers to CFC afterward.

During the focus group sessions, participants reflected on the factors distancing themselves from engaging in Collaborative Fashion Consumption–secondhand, renting, and leasing. Thus, in the present research, it has been found that the main barriers to participating in Collaborative Fashion Consumption for Spanish consumers are:

(A) Product Attributes

1. Hygiene and health concerns: Since clothes are products worn close to the skin, consumers show concerns about their hygiene. Participants' concerns appertain to the overall cleanliness of clothes: bacteria from pre-owners, the possibility of disease transmission, odor, and dirtiness.[2] Participants in FG3 (already SFC) expressed that they do not have these concerns as regular consumers of secondhand fashion, but their families and friends do. Participant 23 narrated:

[2] Concerns about disease transmission through Collaborative Fashion Consumption may have increased due to the Coronavirus crisis, a topic that asks for further research.

Table 5.1 Synthesis of focus group findings

Topic	Focus group 1 (<30)	Focus group 2 (>30)	Focus group 3 (Sustainable consumers)
How often do you buy new clothes?	At least once a month (increasing frequency during sales)	Bimonthly or once in a trimester: when needed or to celebrate special occasions	Mostly, only when needed
Where do you usually buy?	In-store: with friends as a socializing activity/because they like to try clothes on Online: Search in-store and buy online to avoid queues and because online products seem more brand-new than products in-store	50% in-store: to try clothes on 50% online: to avoid crowds and save time	In-store: to check the quality of the fabrics and how they feel on the skin
What is the main reason for you to buy clothes?	They like to wear new outfits all the time To stay trendy, but when they feel that everybody is wearing the same, they do not like it/want it To avoid repeating outfits in pictures (to post on social media)	Out of sheer need To reward themselves	Out of sheer need or when they genuinely like something that they know they are going to wear/keep for a long time
When buying a new garment, which factors do you consider?	Style/price: affordable and on-trend	Quality/price: comfortable and long-lasting at a convenient price point	Quality of materials: natural fibers, comfort and durability In which conditions it has been made
Have you ever bought secondhand garments?	No	Yes, quite often	Very often. Secondhand is their first option

(continued)

Table 5.1 (continued)

Topic	Focus group 1 (<30)	Focus group 2 (>30)	Focus group 3 (Sustainable consumers)
Have you ever rented garments?	No	No	No
Sustainability approach	"For my future self" Interested in sustainable fashion, but not right now. Trying to buy less but better (especially accessories and footwear)	"Now that I know, maybe" They lack knowledge about sustainable fashion, but once they gain some knowledge about it, they are interested. They try to be sustainable consumers by buying in small businesses and neighborhood stores and prioritizing Spanish products	"The clothes you already have are the most sustainable." Buy only when needed, and then prioritize secondhand. Buying brand-new is their last resort Elongate the life span of the clothes they already have by repairing or altering them

> At that time [ten years ago], [buying secondhand clothes] was not well-regarded at all. In fact, one day, I was wearing an adorable sweater, and I told my mother, 'Ah, well, it is secondhand', and she said, 'Take it off, take it off; it is going *to give you* something', and I said 'but mum, there is nothing to worry about'. At that moment, it was very frowned upon [wearing secondhand clothing].

Participants in all focus group sessions reflected on why wearing secondhand garments used to be, and still is today, a hygiene concern. They concluded that it might be related to not knowing who owned the items before. They claimed that, as opposed to secondhand clothing, wearing hand-me-downs is a widespread practice and does not raise the same hygiene concerns; therefore, the prejudice towards secondhand clothing must be related to the fact that the previous owners are strangers and that creates doubts about their possible lifestyle and overall health.

2. Scarce information and lack of trust: perceived price-for-value challenges. Participants raised two points of concern (mainly directed to online secondhand or renting acquisitions):

   (A) Participants were especially skeptical about counterfeits in luxury mid-range goods. They were skeptical that the product they would receive would be the same as they see in the picture while shopping online from both secondhand B2C and P2P sites. Participant 2 claimed:

   > The problem now is that imitations are so well done that, for me, for example, I am not a fashion expert, so I can one hundred percent be fooled.

   (B) Lack of information and the absence of clear and transparent policies regarding the provider's guarantees and the handling of exceptional circumstances, such as damages and customer liability.

3. Lack of ownership: concern towards renting or leasing schemes. Spanish consumers seem not ready to renounce the ownership of their clothes. The sharing economy has flourished in other markets, such as commuting services (*BlaBlaCar*) or accommodation (*Airbnb*). However, since clothing is seen as an extension of the self, it has an additional emotional component that prevents consumers from sharing their clothes. Participants might be willing to rent their clothes and accessories for a one-time-only event, such as weddings, baptisms, graduations, or any other kind of special occasion. Their willingness is based on saving money: usually, for those events where more formal attire is required, they have to buy ad-hoc clothing that, after the event, they are not going to wear again (since repeating outfits for special events is not well regarded); therefore, renting is a way to save money, and precisely because is a one-time-only occasion. They find renting inconvenient, impractical, and not a wise investment for daily life. Participant 8 wondered:

> If clothes are so cheap now, to rent them for your day-to-day, what price would they have? It has to be something even cheaper…

(B) Situational variables

1. Limited locations/selling channels: As pointed out previously, the number of secondhand stores and renting services in Spain, at the moment, is scarce, mainly out of the big cities, so potential consumers have to make an extra time effort to find what they need. Fast-fashion stores occupy the best locations in cities, whereas secondhand stores are scarce, requiring additional searching and commuting time.
2. Product selection/availability: Participants complained that searching online and in-store in secondhand outlets and renting services usually takes much more time to find items in their style and size. The product selection/availability is much more restricted than in traditional shopping outlets. In Spain, secondhand stores usually belong to NGOs (such as *Humana*, for instance). For that reason, the stores generally differ considerably from conventional fast-fashion spaces in their ambient distribution and clothes disposition. Consumers believe that secondhand stores are poorly organized, leading to frustrating shopping experiences. Searching online for secondhand items or renting is also perceived as a much more time-consuming task than searching on conventional brands' websites since the product selection is more limited (regarding style) and availability, regarding sizes, is scarce (most times both in renting services and secondhand outlets the number of garments per size is very limited or even just one).

(C) Individual traits

1. Consumption habits: Spanish consumers are used to low-price apparel from fast-fashion brands, and renting or leasing their clothes would dramatically

alter their consumption patterns. Fashion shopping is a leisure activity, a way to fill out their spare time and socialize with friends (as previously highlighted, consumers create rituals of going shopping or going to the sales with their friends, particularly younger consumers). Moreover, in the three focus group sessions, consumers stated that they hate to return items for time costs (such as scheduling appointments with the delivery services) and emotional reasons (detachment), so renting or short-term leasing is not well regarded. Thus, as already pointed out, renting is only seen as an option for specific events, such as weddings, or for transition periods in life, like maternity or children in the process of growing.

2. Lack of knowledge: Renting is still a fledgling business in Spain, and many consumers are unaware of which companies provide this service and its terms and conditions. They have the same problem with secondhand. They may know some traditional secondhand stores, but they do not acknowledge online websites providing this service B2C.
3. Self-expression: Consumers worry about the self-image and social image they might project, mainly by buying secondhand clothing. Consumers of secondhand clothing stated that they hide deliberately from their friends and family to buy secondhand apparel because they feel that they would be labeled "poor" or "stingy". They also stated that if a coworker compliments their outfit and it is indeed a secondhand item, they will avoid telling her/him. Participant 12 explained:

> If you are one day dressed in your normal clothes and you say to someone, 'I bought this secondhand,' they already see you with different eyes, don't they? If you say nothing, no one knows if you bought your clothes precisely at the store next door.

In Spain, there still exists a stigma around buying secondhand clothing, on the one hand, for hygiene reasons, and on the other hand, because it is associated with a lower social status. However, the same stigma does not exist for hands-me-down or buying secondhand from other markets, such as London or Paris (were explicitly mentioned). Whereas secondhand in Spain does not have a good reputation, apparel from more consolidated secondhand markets has a certain prestige. Participant 5 claimed:

> In Spain, I think there is no such culture [secondhand]. There are no secondhand stores where the garments look good, like well cared for, that they have been dry-cleaned, that the linings are intact, or even that they have fixed them themselves before selling them, whereas in other countries there is much more [secondhand outlets that sell well-cared garments].

Participant 12 commented in the same vein,

> In Spain, there has never been a secondhand culture. So, if people from Spain went abroad, there were people who went to London on weekends to buy clothes, and maybe they bought them at the flea market, and they came back with the clothes since they were from London, they were already more beautiful clothes, you know? So, those clothes had like a certain cache, you know? Here in Spain, if you bought something secondhand, you were like a poor person, you know? To put it in some way.

This could be because most secondhand stores in Spain, as stated before, belong to NGOs. Lastly, younger consumers refuse to buy in the secondhand market because they believe it lacks trendiness. Conversely, trendiness is a factor they appreciate in leasing and renting schemes.

Lastly, Fig. 5.3 summarizes the identified barriers for participating in Collaborative Fashion Consumption categorized by individual traits, product attributes and situational variables.

Conversely, this research has found that the main drivers for Spanish consumers to engage in Collaborative Fashion Consumption are:

1. Sustainability motivations: Secondhand is the preferred alternative for sustainable consumers because they can find various styles, qualities, and sizes (whereas in sustainable fashion stores, they feel more limited regarding style and plus-sizes). Especially working-appropriate clothes that, in their opinion, no sustainable brand offers. They believe that their decision to engage with alternative consumption channels demonstrates their resourcefulness and dedication to

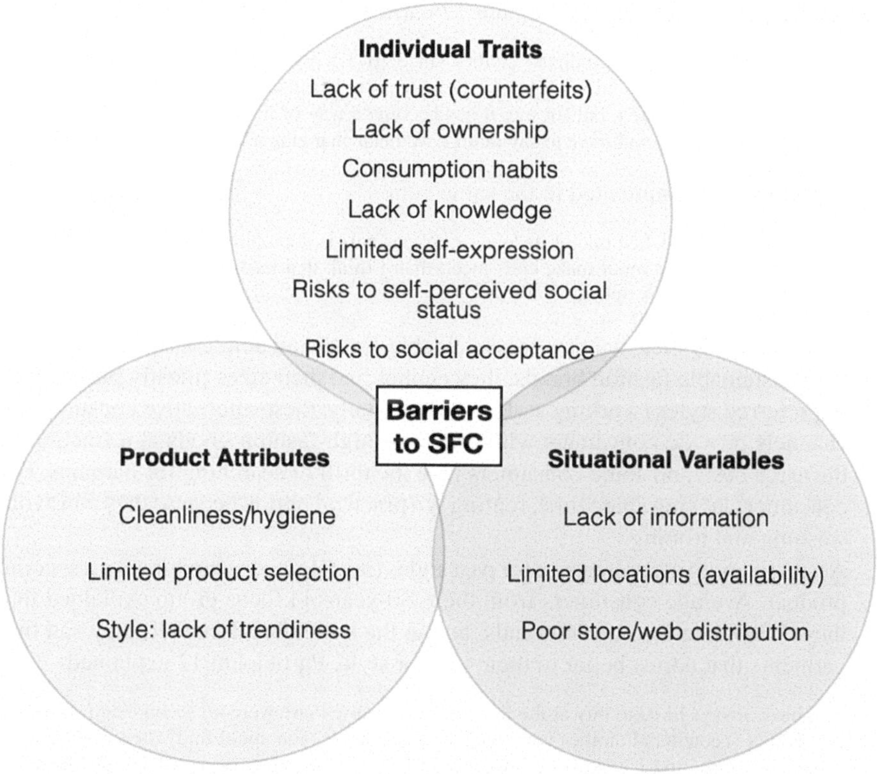

Fig. 5.3 Synthesis of the main barriers to participating in collaborative fashion consumption. *Source* Author

reducing wasteful practices. Moreover, they know that once they finish their items, they can resell them again in the secondhand market, thus closing the loop. The quote from Orsola de Castro, "The most sustainable garment is the one already in your wardrobe", is their motto so that they will prioritize already-made garments over brand-new ones. Participant 18 affirmed:

> The most sustainable clothing is what you already have right now, the most sustainable, what has already been produced, the clothes you do not throw away, even if you bought it at a fast-fashion store. You must have them for years and years and years.

They consistently question personal needs and focus on purchasing clothing only when there is a genuine need. Participant 12 expressed:

> I think we have to sustain things differently; we cannot consume this much, and we have to take care of the planet and reverse everything we have been doing wrong.

2. Utilitarian motives: Secondhand is valued as a good alternative for saving money. In fact, some focus group participants stated that they became first in touch with secondhand amidst a financial crisis. They were skeptical initially, but since they did not have other options, they started buying secondhand, and now they have adopted it as a lifestyle. Participant 12 narrated:

   > I have been buying secondhand clothes since 2011 when I started to buy them quite often, and I have to confess, it was out of necessity because, at that time in my life, it was a bit complicated, but for me, it has become a way of living, because it seems to me that it is great, so I have to say that I have been changing a lot throughout my life.

   Participant 10 commented in the same vein:

   > I also think that when one of us [she or her partner] is unemployed, and the money is not enough, you cannot make ends meet, then I think that leads us to the secondhand market when one is tighter.

   Secondhand is their way to go sustainable for sustainable consumers because, with sustainable fashion brands, they cannot find their sizes (mostly plus sizing) or preferred styles (working styles). Additionally, these alternative consumption channels provide consumers with access to high-fashion goods at a fraction of the usual cost, and some consumers love the thrill of searching for bargains. For consumers lacking spare time, renting is a practical option because they can avoid washing and ironing.
3. Aesthetic motives: preference for past styles (nostalgia) or searching for a specific product. Average consumers from the > 30-year-old focus group explained that they do not like the current trends, and in the secondhand market, they can find garments that adjust better to their taste or style. Participant 12 explained:

   > I have always liked to buy at the flea market because there were no secondhand stores yet, but secondhand clothes were sold in flea markets, you could find when you went on vacations, and I liked it because a lot of the clothes came from outside of Spain, so perhaps they were American, British, … and I liked to buy them because they were different.

Others go to the secondhand market searching for a specific product from a past season that they liked but missed the opportunity to buy. Participant 11 explained:

> I have also bought secondhand garments no longer available in stores, clothes that I could not buy first-hand, and I was left wanting them, and I have found them secondhand.

Sustainable consumers like the idea of not being uniform with the 'mass' of consumers; hence, buying secondhand enables them to define their personal style in opposition to the traditional market's mass trends. Moreover, they believe fashion consumers have become 'prisoners' of the big fashion group dictates. Sustainable fashion and secondhand or vintage clothing are the instruments to regain the power to express themselves with their clothes, thus enabling aesthetic creativity. Participant 20 reflected:

> I believe that people should regain the ability to express themselves, not that a brand decides how we have to dress.

4. Hedonic motives: Consumers are thrilled to hunt for bargains (especially among luxury products) and search for rare items that would otherwise be unattainable. Participant 13 narrated:

   > In stores like *Humana*, I have had great surprises because I have found jeans with the label, new, I have found sweaters in perfect condition, without any signs of wear, and each of one for one euro!

   Whereas Participant 14 claimed:

   > Maybe you do not have something from a particular brand because you cannot afford it, and you go there [to the secondhand store], and you can get it at a good price and in good condition.

   In the same way, renting might trigger hedonic values because it allows consumers to try new styles without the financial strain of traditional ownership.

Lastly, Fig. 5.4 summarizes the identified drivers for participating in Collaborative Fashion Consumption categorized by individual traits, product attributes and situational variables.

5.5 Discussion

Early environmental discussions emphasized the need for prudent resource management to prevent the depletion of finite resources. While concerns about immediate or short-term material scarcity have subsided, the environment's capacity to handle material inputs without detrimental effects remains limited. The evolving debate on resource productivity reflects a growing desire to achieve a balance between economic and environmental goals (efficiency objective) and address the ethical

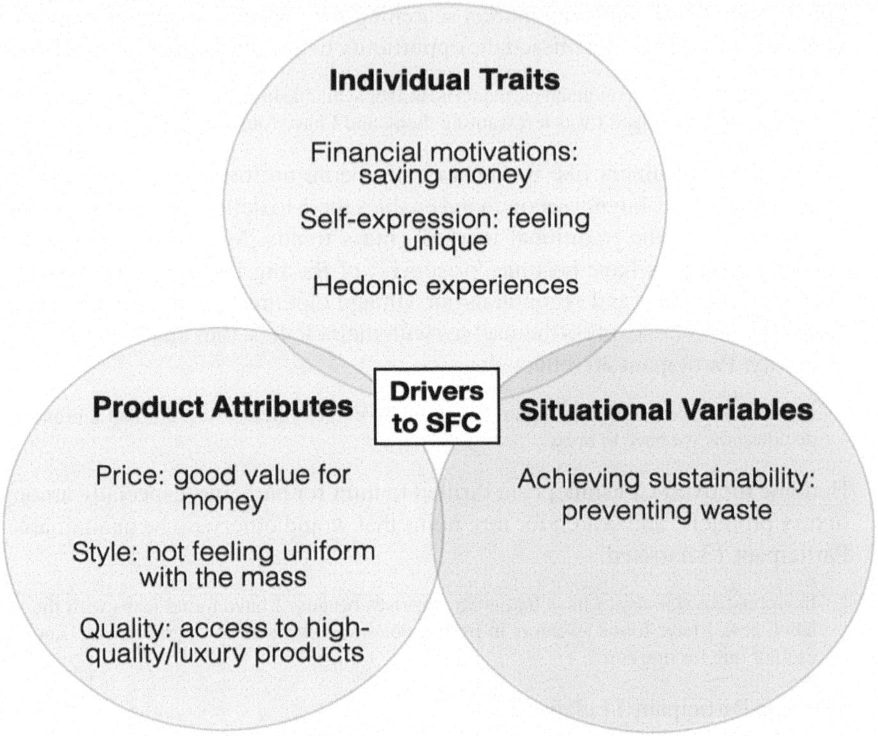

Fig. 5.4 Synthesis of the main drivers to participating in collaborative fashion consumption. *Source* Author

implications of excessive consumption in developed nations, particularly in relation to less industrialized regions and future generations (moral objective) (Cooper, 2005).

Therefore, involvement from governments, businesses and consumer behavior and lifestyle are key factors to transition towards a more sustainable world, as posited by SDG12.

5.5.1 Implications for Managers, Practitioners, and Decision Makers

Against this background, I will list a series of recommendations for governments, businesses, and consumers to better conform to the principles of SDG12.

For instance, governments' future awareness and education campaigns could concentrate on the fashion industry's impacts, for example, pointing out how the fashion industry is an asset to the Spanish economy and why it is essential to preserve it and prioritize national production or remarking on the cultural value and traditions of the Spanish fashion sector.

Fashion shopping nowadays is more a leisure and social activity than something done out of necessity. To preserve consumers' enjoyment while lowering sustainability impacts, local town councils could provide spaces for alternative leisure and social activities (alternative to shopping) among younger generations—for example, spaces where swap parties could be carried out. Swap parties are events where participants bring their clothes and accessories to trade them with other participants, and they are standard practice in countries such as the UK or the Netherlands. The parties could be animated with live music and have a low entrance fee to promote participation and generate revenue. This kind of event would provide participants with entertainment and social opportunities alternative to conventional shopping–promoting clothes circularity–and, therefore, represent a more sustainable approach to fashion consumption.

Another action governments could provide or incentivize to increase fashion product circularity is the creation of fashion libraries. Fashion libraries are garment depositories where consumers can borrow clothes through a subscription or monthly/yearly fee. These have been proven successful, for example, in The Netherlands or Sweden, especially for children's wear or maternity clothes (European Circular Economy Stakeholder Platform, 2019), since these are transitory periods in life. Usually, clothes have to be bought ad hoc for these stages and soon become no longer useful.

Fashion companies seeking innovative and sustainable business models could embrace collaborative consumption as a promising avenue to enhance efficiency and sufficiency (Kant Hvass, 2015). Firstly, they could introduce a dedicated in-store exchange program where customers can trade in their used items for discounts on new products (as *Zara* or *H&M* already do); secondly, they could develop their own platforms for repurposing and monetizing pre-owned fashion items. For instance, *Farfetch* resales used-luxury fashion goods under the category 'pre-loved,' and *Eileen Fisher* offers secondhand products at *Green Eileen.* Similarly, to committing to circularity, fashion companies could provide altering and repair services for extended periods after acquisition. Altering garments or accessories could be presented as a form of customization that would target consumers' self-enhancement values (self-expression and individualism) and self-transcendent (sustainability by keeping garments longer).

Secondhand and renting outlets should use visual merchandising strategies to improve in-store and online garment displays to appear more like traditional retail outlets and thus more inviting. They could provide editorial style trends to make products more appetizing to consumers. In addition, they could create narratives as 'treasure hunting' to appeal to consumers' hedonistic values and employ storytelling to, for example, narrate the history of the previous garment owner (respecting her/his privacy) and thus surpassing the barrier of lack of trust in secondhand products related to not knowing who owned it before. To reach new customers, secondhand and renting outlets could increase their visibility by opening pop-up stores in strategic locations (maybe with government subsidies) and celebrating clothes-sharing events (swap parties).

To conclude, consumers can change their habits, from the pre-purchasing phase to the use and disposition of their garments to be more sustainable. For instance, although shopping nowadays is done more as a leisure and social activity than out of need, consumers could try to engage more often in Collaborative Fashion Consumption (as the swap parties explained above, or secondhand and renting) and to control impulse buying (for example, by assigning a monthly budget to fashion consumption). These practices could lead not only to more sustainable habits but also to economic savings and avoid post-shopping regrets or cognitive dissonance[3]; therefore, potentiating both self-transcendent and self-enhancement values.

Moreover, more trend-prone consumers could consider renting their clothes, which would allow them to have a high product rotation at a lower cost and minimize clothing waste. When clothes are no longer wanted, consumers should consider selling them secondhand. If reselling is not an option, consumers could consider exchanging them with friends/family members, donating or adequately disposing of garments for them to be recycled.

5.5.2 Future Research Agenda and Conclusions

Pathways for future research could be to perform a similar analysis in other countries to determine how the cultural and social context influence fashion consumption since these appear to be critical factors in sustainable fashion purchasing behavior.

Since this research is solely focused on the consumers' perspective, future research may concentrate on how fashion brands, renting and secondhand outlets disseminate and communicate their sustainability initiatives and/or the volume and frame of sustainability content in mass media outlets to have a broader perspective of the state of the sustainable fashion market in Spain.

In conclusion, transitioning to a sustainable fashion industry requires a concerted effort from all stakeholders. By working together, governments, businesses, and consumers can foster a more environmentally and socially responsible fashion industry for the future.

[3] According to Lee and Li (2013), *cognitive dissonance* "can be divided into cognitive and emotional perspectives. From a cognitive point of view, the dissonance is the difference in consumer's belief after making a purchasing decision. In contrast, from an emotional point of view, the dissonance is the uncomfortable feeling following a purchasing decision" (p. 72). Thus, dissonance is the satisfaction or regret after a purchase. Generally, regret follows impulse buying, while satisfaction is the aftermath of planned acquisitions.

References

Bardhi, F., & Eckhardt, G. M. (2012). Access-based consumption: The case of car sharing. *Journal of Consumer Research, 39*(4), 881–898. https://doi.org/10.1086/666376

Becker-Leifhold, C., & Iran, S. (2018). Collaborative fashion consumption–drivers, barriers and future pathway. *Journal of Fashion Marketing and Management, 22*(2), 189–208. https://doi.org/10.1108/JFMM-10-2017-0109

Belk, R. W. (2014). You are what you can access: Sharing and collaborative consumption online. *Journal of Business Research, 67*(8), 1595–1600. https://doi.org/10.1016/j.jbusres.2013.10.001

Bocken, N. M. P., Short, S. W., Rana, P., & Evans, S. (2014). A literature and practice review to develop sustainable business model archetypes. *Journal of Cleaner Production, 65*, 42–56. https://doi.org/10.1016/j.jclepro.2013.11.039

Boons, F., & Ludeke-Freund, F. (2013). Business models for sustainable innovation: State-of-the-art and steps towards a research agenda. *Journal of Cleaner Production, 45*, 9–19. https://doi.org/10.1016/j.jclepro.2012.07.007

Botsman, R., & Rogers, R. (2011). *What's mine is yours: How collaborative consumption is changing the way we live*. Collins.

Bras-Klapwijk, R. M., Marjolijn, J., & Knot, C. (2001). Strategic environmental assessment for sustainable households in 2050: Illustrated for clothing. *Sustainable Development, 9*(2), 109–118. https://doi.org/10.1002/sd.157

Business of Fashion. (2021). *Why Kering invested in Vestiare Collective*. Retrieved February 19, 2022, from https://www.businessoffashion.com/articles/luxury/kering-acquires-5-stake-in-vestiaire-collective/?fbclid=IwAR24aqczOFk6hErMrGIcnEmJEb1qaqcb0mAkpi84YUq-1EfHWCK-NEOFp8A

Business of Fashion. (2022a). *Why Shein might be worth 100$ Billion, in four charts*. Retrieved April 19, 2022, from https://www.businessoffashion.com/articles/retail/why-shein-might-be-worth-100-billion-in-four-charts/

Business of Fashion. (2022b). *How to compete with Shein*. Retrieved April 19, 2022, from https://www.businessoffashion.com/articles/retail/how-to-compete-with-shein/

Canalejo, A. (2020). 5 plataformas de alquiler de moda para estrenar tendencias cada mes sin arruinarte. *Elle*. Retrieved November 11, 2021, from https://www.elle.com/es/moda/compras-elle/a34625691/alquiler-ropa-mujer-mejores-plataformas/

Caro, F., & Martinez-de-Albeniz, V. (2015). Fast fashion: Business model overview and research opportunities. In N. Agrawal & S. Smith (Eds.), *Retail supply chain management: Quantitative models and empirical studies* (pp. 237–264). Springer.

Cooper, T. (2005). Slower consumption. Reflections on product life spans and the throw-away society. *Journal of Industrial Ecology, 9*(1–2), 51–67. https://doi.org/10.1162/1088198054084671

Cyr, J. (2019). *Focus Groups for the social science researcher*. Cambridge University Press. https://doi.org/10.1017/9781316987124

Dahlbo, H., Aalto, K., Eskelinen, H., & Salmenpera, H. (2017). Increasing textile circulation–consequences and requirements. *Sustainable Production and Consumption, 9*, 44–57. https://doi.org/10.1016/j.spc.2016.06.005

European Circular Economy Stakeholder Platform. (2019). *Own less live more*. (10 Oct 2019). https://circulareconomy.europa.eu/platform/en/good-practices/own-less-live-more-just-borrow-fashion-library

European Environment Agency. (2019). *Textiles in Europe's circular economy*. Retrieved February 19, 2022, from https://www.eea.europa.eu/publications/textiles-in-europes-circular-economy

Fletcher, K. (2008). *Sustainable fashion and textiles. Design journeys*. Earthscan.

Fletcher, K. (2016). *Craft of use: Post-growth fashion*. Routledge.

Global Living Wage Coalition. (2021). *Giving workers a decent standard of living*. Retrieved October 10, 2021, from https://www.isealalliance.org/about-iseal/our-work/global-living-wage-coalition

ILO. (2015). *Minimum wages in the global garment industry.* Retrieved October 10, 2021, from https://www.ilo.org/asia/publications/WCMS_436867/lang--en/index.htm

International Encyclopedia of the Social & Behavioral Sciences. (2001). https://www.sciencedirect.com/referencework/9780080430768/international-encyclopedia-of-the-social-and-behavioral-sciences

Iran, S., & Schrader, U. (2017). Collaborative fashion consumption and its environmental effects. *Journal of Fashion Marketing and Management, 21*(4), 468–482. https://doi.org/10.1108/JFMM-09-2016-0086

Jackson, T., & Shaw, D. (2008). *Mastering fashion marketing.* Palgrave Macmillan.

Kant Hvass, K. (2015). Business model innovation through second-hand retailing–a fashion industry case. *The Journal of Corporate Citizenship, 57*, 11–32. https://doi.org/10.9774/GLEAF.4700.2015.ma.00004

Lee, L.Y., & Li, L.Y. (2013). A study on effect of cognitive dissonance on consumption value, anticipated satisfaction and word-of-mouth communication. *The Journal of Global Business Management, 9*(2).

Marshik, C. (2011). Smart clothes at low prices. In I. Parkins & E. M. Sheehan (Eds.), *Cultures of femininity in modern fashion.* University of New Hampshire Press.

Mayring, P. (2010) *Qualitative Inhaltsanalyse–Grundlagen und Techniken.* Beltz Verlag.

Micolet. (2020). *Informe de la moda circular 2020.* Retrieved November 11, 2021, from https://www.micolet.com/informe-moda-circular-2020

Mohr, L. A., Webb, D. J., & Harris, K. E. (2001). Do consumers expect companies to be socially responsible? The impact of corporate social responsibility on buying behavior. *Journal of Consumer Affairs, 35*(1), 45–72. https://doi.org/10.1111/j.1745-6606.2001.tb00102.x

Niinimäki, K., Peters, G., Dahlbo, H., Perry, P., Rissanen, T., & Gwilt, A. (2020). The environmental price of fast fashion. *Nature Reviews, 1*, 189–200. https://doi.org/10.1038/s43017-020-0039-9

Pencarelli, T., Ali Taha, V., Skerhákova, V., Valetntiny, T., & Fedorko, R. (2020). Luxury products and sustainability issues from the perspective of young Italian consumers. *Sustainability, 12*, 245. https://doi.org/10.3390/su12010245

Peters, G. M., Sandin, G., & Spark, B. (2019). Environmental prospects for mixed textile recycling in Sweden. *ACS Sustainable Chemical Engineering, 7*, 11682–11690. https://doi.org/10.1021/acssuschemeng.9b01742

Retema (2021). *La compra de segunda mano aumenta el 38% en España en los últimos años.* (17 May 2021). Retrieved November 11, 2021, from https://www.retema.es/noticia/la-compra-de-ropa-de-segunda-mano-aumenta-el-38-en-espana-en-los-ultimos-anos-dsb

Reyes, J., Román, R., Sanz-Díaz, T., & Mundaca, L. (2020). *Collaborative fashion consumption in Sweden and Spain: market, environmental and behavioural insights.* https://lup.lub.lu.se/search/files/81060034/BE_USE_Project_Report_Phase_1_CFC_FINAL.pdf

Sajn, N. (2019). *Environmental impact of the textile and clothing industry: What consumers need to know.* European Union. European Parliamentary Research Service. http://www.europarl.europa.eu/thinktank/en/document.html?reference=EPRS_BRI%282019%29633143.

Sandin, G., & Peters, G. M. (2018). Environmental impact of textile reuse and recycling–A review. *Journal of Cleaner Production, 184*, 353–365. https://doi.org/10.1016/j.jclepro.2018.02.266

Schmidt, A., Poulsen, P. B., Watson, D., Roos, S., & Askham, C. (2016). *Gaining benefits from discarded textiles: LCA of different treatment pathways.* Retrieved February 10, 2022, from https://norden.diva-portal.org/smash/get/diva2:957517/FULLTEXT02.pdf

Schor, J.B., & Fitzmaurice, C.J. (2015). Collaborating and connecting: the emergence of the sharing economy. In L. Reisch & J. Thogersen (Eds.), *Handbook of research on sustainable consumption* (pp. 410–425).

Schrader, U. (2001). *Konsumentenakzeptanz Eigentumsersetzender Dienstleistungen.* Lang.

Thredup (2021). *2021 resale report.* Retrieved February 19, 2022, from https://www.thredup.com/resale/#resale-industry

United Nations. (2019). Sustainable Development Goals Report 2019. Retrieved January 24, 2024, from chrome-extension://efaidnbmnnnibpcajpcglclefindmkaj/https://unstats.un.org/sdgs/report/2019/The-Sustainable-Development-Goals-Report-2019.pdf

United Nations Sustainable Development Goals. (2015). *Resolution A/RES/70/1. Transforming our world: the 2030 Agenda for sustainable development.* Retrieved June 15, 2021, from https://www.un.org/ga/search/view_doc.asp?symbol=A/RES/70/1&Lang=E

Waight, E. (2013). Eco babies: Reducing a parent's ecological footprint with second-hand consumer goods. *International Journal of Green Economics, 7*(2), 197–211. https://doi.org/10.1504/IJGE.2013.057444

Woolridge, A. C., Ward, G. D., Phillips, P. S., & Collins, M. (2006). Life cycle assessment for reuse/recycling of donated waste textiles compared to use of virgin material: An UK energy saving perspective. *Resource, Conservation and Recycling, 46*, 94–103. https://doi.org/10.1016/j.resconrec.2005.06.006

Chapter 6
"The Fashion Industry in Crowdfunding Sites: An Analysis of the Sustainable Consumption Cues that Shape the Future of the Fashion Business"

Ana M. Gómez-Olmedo and María Eizaguirre Diéguez

6.1 Introduction

While evidence shows how the textile sector is the second most pollutant industry, with 10% of global carbon emissions, the depletion of natural resources is reaching unprecedented levels given that the demand for fashion keeps increasing. A single cotton t-shirt takes 2.5 years of a person's drinking water to produce. Compared to 2000, apparel purchases have increased by 60% (Niinimäki et al., 2020). However, 40% of clothes in our wardrobes will never be worn (European Parliament, 2023). If these increasing trends continue, the demand for the natural resources required to keep this growth will triple by 2050. Given these facts and projections, undoubtedly, the textile industry is a major player in the transition to sustainability in business. Additionally, this prominent impact of the textile industry and fashion business in sustainable development brings a major responsibility to mitigate the damage.

Thus, the fashion business has been actively pursuing its transformation towards sustainability and SDGs might serve as both a mandate and a lever to pursue the complex challenges that go along with it. While SDG12: Responsible Consumption and Production might be seen as the most straightforward objective, it is just a gateway to the other SDGs given the nature of the textile industry value chains. To name a few, not only is the environment suffering damage (SDG13), but also the inequalities (SDG10) and social injustice (SDG16) that come as a result of the loss of biodiversity and climate catastrophes. Additionally, increasing levels of poverty (SDG1) and

A. M. Gómez-Olmedo (✉) · M. Eizaguirre Diéguez
Department of Marketing, ESIC University, Madrid, Spain
e-mail: ana.gomez@esic.university

Department of Marketing, ESIC Business and Marketing School, Madrid, Spain

A. M. Gómez-Olmedo
ESIC University, Avenida de Valdenigrales s/n, 28223 Pozuelo de Alarcón, Madrid, Spain

J. F. Gallego-Nicholls and A. Carrilero-Castillo (eds.), *Textile Industry and the SDGs*, SDGs and Textiles, https://doi.org/10.1007/978-981-97-5073-3_6

unfair employment (SDG8) worsen population well-being (SDG3) (Cai & Choi, 2020).

Global sustainability programmes are put forward by large corporations such as Walmart, Patagonia or Inditex (Del Pilar Quiroz Dzhengiz et al., 2023; Galvan et al., 2021). These programmes capture the complexity that the industry transformation entails. From a reshape of their whole value chain to the education of their consumers so that their purchases are informed by the social, environmental, and economic practices that are required to achieve the objectives pursued. To do so, circular economy strategies are offered, such as the framework put forward by Morseletto (2020) which includes 10 circular economy strategies that are comprised of three main targets, namely, smart product usage and manufacture, the extended lifespan of products and their parts, and useful application of materials.

For traditional corporations, the sustainability transition might require large investments and a powerful will from the people behind it. However, entrepreneurs in the fashion business might most easily bring some disruptive ideas that could help to accelerate or at least push forward the changes needed (Moon, 2018). The most effective process and strategies for this transformation, from big corporations to start-ups, constitute an encouraging avenue for theoretical and empirical research (Nilsson & Göransson, 2021).

While acknowledging the complexities of this transformation, for the industry to favour more sustainable businesses, collective action seems necessary. In this collaborative thinking, co-creation initiatives, involving different stakeholders, might serve as part of the solution. In this, the entrepreneurs' ecosystems that allow more room for innovation might help to redesign business models and value chains while promoting global awareness and social support (Di Vaio et al., 2022).

Within co-creation initiatives, crowdfunding platforms are seen as experimental settings where entrepreneurs have the chance to test their initiatives before opening their business solutions to global markets. In doing so, social capital is built, and financial support is achieved (Böckel et al., 2021). Thanks to this, the knowledge captured in these platforms is highly valuable. Innovation in fashion ideas as well as social capital strength can be assessed by looking at which ideas are financially backed up and the business ideas that are behind crowdfunding campaigns. It also serves to identify future trends so that industry projections can be identified as well as major concerns that drive innovation based on underserved consumers' needs (Deng et al., 2022).

While it is commonly agreed that corporations in the fashion industry have traditionally set the agenda of consumer's behaviour, given as a result unsustainable behaviours such as the fast-fashion movement, it has also helped to democratize the access to apparel while empowering consumers' identity (Dzhengiz et al., 2023).

Notwithstanding the conflict that arises in the promotion of these behaviours, the great influence of these corporations could serve as keen educational campaigns that shape consumers' behaviours towards a more mindful consumption (Sheth et al., 2011). To this end, circular economy initiatives have emerged as a facilitator to a more sustainable fashion business.

In this joined-up transition towards sustainability, new business ideas should presumably be built under the circular economy's premises, as if these ideas were the only conceivable ones. Thus, from their roots, they promote a more mindful consumption by offering alternative modes of consumption as new business ideas (Mora-Cruz & Palos-Sanchez, 2023). To do so, entrepreneurs as well as the platforms that help to disseminate more responsible consumption alternatives should be assessed so that efforts are ensured towards a value chain closely aligned to the common Agenda 2030.

This research works towards the aim of addressing these challenges. By conducting a thematic analysis of crowdfunding projects in one of the major platforms, Kickstarter, the main drivers of innovation in the fashion industry are identified to evaluate whether they nurture mindful consumption behaviour. In doing so, the main themes, trends and sustainable cues are identified. These aspects are discussed so that an interpretation of the results provides guidance for future research as well as a strategic analysis for managers in the field. By doing so, this research is aligned with recent calls about the readiness of the crowdfunding literature to contribute to sector-specific discussions while working towards common goals such as the SDGs (Dalla Chiesa et al., 2022).

To address the objectives proposed, this chapter is structured as follows. First, it offers a literature review about the field of fashion, sustainability and co-creation. This serves to build the theoretical corpus in which later analysis is grounded. Secondly, a thematic analysis is conducted where four main categories emerge, namely, *accessories, underwear, apparel, footwear*. The results are then offered with an interpretation of those results with theoretical and managerial implications and concluding remarks.

6.2 Sustainable Fashion Entrepreneurship

Excessive consumption practices exerted from a variety of stakeholders, businesses, institutions, entrepreneurs, or consumers, are threatening the well-being of our ecosystems. This growing and excessive demand requires the depletion of a larger number of resources, most of which are not brought back to the production process. If the pace of these trends continues, it is projected that we will need two or three planets to fulfil the needs of this vicious circle reaching a tipping point by 2030 (Pero et al., 2020; Wren, 2022). Acknowledging that consumers should not take full responsibility for these consequences, the role of entrepreneurs, among the other stakeholders, in offering solutions for a more sustainable fashion industry is deemed critical (Fan & Chang, 2023).

In a review of recent academic literature about the role of entrepreneurship in the promotion of sustainable fashion business models, the role of this industry in the environment is commonly recognized (Heidenreich et al., 2022; Jung & Kim, 2023; Ostermann et al., 2021). Social and environmental issues derived from new products and services are seen as both strategic concerns by companies as well as the ground in which to build their business (Ostermann et al., 2021). Therefore, sustainability is at the core of new business ideas, strategies and processes and should guide decision-making in all the stages of the value chain.

In this, businesses and entrepreneurs, in particular, also take responsibility for the promotion of social and environmental consciousness in purchases so that consumers shift their behaviours to more sustainable options. However, to add complexity, it is also acknowledged that paradoxically, consumers' sustainable intentions are not always coherent with actual purchases. Moreover, in the field of fashion, consumers' demand for sustainable business models are not yet considered a strong driver for companies to shift unsustainable industry practices (Ostermann et al., 2021). To this end, and from the perspective of the demand side, marketing efforts to guide sustainable consumption purchases are paramount (Jung & Kim, 2023). To do so, the identification of consumer demands to build active interaction between entrepreneurial ideas and their intended customers is an essential task. In this, crowdfunding platforms are crucial (Lipusch et al., 2020).

Crowdfunding platforms are invaluable tools for this purpose, promoting the transition to sustainable new business ideas, given that they are seen as experimental settings to customize products and services to shape future demand (Böckel et al., 2021). Additionally, the emphasis on sustainable products, branding and consumption are important for consumers to feel a sense of belonging to a wider sustainability-aware community that will help them in moments of paradoxically irresponsible behaviour. Moreover, crowdfunding platforms allow entrepreneurs to build long-lasting relationships due to a sense of psychological ownership of those that financially support the ideas, i.e. the backers. Backers might end up turning into customers whose connection with the brand is stronger due to this long-term interaction (Belleflamme et al., 2014; Mollick, 2014). These relationships are an invaluable tool to nurture co-creation within a new mindset, one of mindful consumption.

Additionally, in the field of fashion, previous literature has shown how marketing activities should explore further messages rather than just the "eco-friendly" labelling. Material sources, processes, or benefits could be added as requirements for sustainable business ideas. Adding to these and considering that most crowdfunding ideas require business model innovation and the design of new customer experience journeys, the use of messages that emphasize the benefits of alternative modes of consumption as well as the number of backers supporting these ideas are seen as important (Heidenreich et al., 2022). All in all, crowdfunding platforms show high potential to promote more mindful and responsible consumption in a field dominated by fast fashion business models.

6.3 Mindful Consumption in the Fashion Industry

In the promotion of more sustainable fashion products and services, mindful consumption has some potential benefits. Mindful consumption is defined as a "temperance in acquisitive, repetitive and aspirational consumption at the behaviour level, ensuing from and reinforced by a mindset that reflects a sense of caring toward self, community, and nature" (Sheth et al., 2011, p.30). Additionally, this temperance could be enacted in three types of consumption behaviours, namely, acquisitive, repetitive, and aspirational consumption. These consumption behaviours refer to consuming more than one is capable of using (acquisitive), impulsively (repetitive) and ambitiously (aspirational) (Gupta & Verma, 2019; Mahmud et al., 2019). Although all of them seem relevant to drive the aim for a more sustainable fashion industry, for fast fashion, repetitive consumption behaviour, that is, the prevalence of compulsive buying, is seen as a major concern (Del Prete, 2022; Papasolomou et al., 2023).

Acknowledging these opportunities and challenges, mindful consumption could be seen as an enabler of an enhanced awareness to shift from irresponsible purchases in a field, fashion, where awareness is deemed as key (Areias et al., 2023). Moreover, conscious fashion purchases are labelled as "ethical fashion" where consumers and fashion supply change stakeholders play their part. In this, product attributes and alternative choices are seen as higher predictors of mindful fashion consumption behaviours (Ray & Nayak, 2023). More particularly, mindful consumption behaviours could be cultivated by providing clear social and environmental messages complemented with new technologies such as chatbots. Together, mindful consumption practices in crowdfunding settings have the potential to build greater awareness and stronger connections among crowdfunding communities. Moreover, the promising potential of building awareness through crowdfunding platforms is closely related to financial support behaviour. Thus, the basis of mindful consumption, an awareness linked with a behaviour, can be fully leveraged by working in crowdfunding settings.

To reach this potential, it should be acknowledged that previous research shows how a lack of accuracy in conveying sustainability impacts, regarding supply chains, materials or circular economy related aspect (e.g., recycling) holds back any effort to shift consumers' behaviours (Papasolomou et al., 2023). In contrast, the benefits of greater connection between social and environmental impacts and our consumption behaviour have been empirically evidenced. Thus, greater connection, that is, an enhanced awareness of the impact of our consumption in ourselves, our communities, and the environment, drives motivation and commitment regardless of more functional attributes such as money.

Additionally, this commitment is achieved long-term as opposed to the ephemeral moments of joy that fast fashion consumption delivers (Areias et al., 2023). Not surprisingly, a novel concept has emerged, namely mindful clothing consumption, where individuals' social and environmental concerns are taken into account when acquiring, using and disposing of clothing items (Armstrong, 2021); it is a concept and a lifestyle that has seen a promising line of research (Aleem et al., 2022).

6.4 Crowdfunding Success Factors in Sustainable Fashion

Previous literature has identified key factors for the success of crowdfunding initiatives. These factors should be seen through the lens of sustainability and, in particular, from the field of fashion (Cricelli et al., 2022; Dalla Chiesa et al., 2022; Deng et al., 2022; Mora-Cruz & Palos-Sanchez, 2023). Out of the previous elements identified in past literature (e.g., environmental attributes, campaign information, creators, social interaction (e.g., comments), alliances, number of supporters), social capital shows a major influence in the success of crowdfunding initiatives; messages regarding safety attributes are also seen as important (Deng et al., 2022). Social capital encompasses major factors that might best describe sustainable fashion initiatives, information about product design, manufacturing, marketing messages and the context of consumption which could be seen as essential factors to bring the support of backers for these initiatives (Zhang et al., 2022).

It is important to note that early-stage innovation ideas seem to receive greater support from backers. This brings support to the notion that working towards sustainable fashion business should not only relate to bringing to market responsible products and services but also to build greater community connection (Lipusch et al., 2020). Thus, crowdfunding projects seem to represent a place in the community for the backers that support the initiatives, a sense of belonging to a greater purpose with the emotional connection that this facilitates to brands. Crowdfunding projects with lower goals but with a level of quality guaranteed also allow a sense of short-term accomplishment which seems to be appealing to backers. Environmental messages rather than social narratives seem also to drive greater support. This should be interpreted with caution given that the discussion around fashion is mainly environmentally centred. The aspirational consumption tendency is also observed by the fact that distinction seems to be associated with higher incentives for customers (Dalla Chiesa et al., 2022). In summary, highly incentivized environmental products and services, at their early stage of innovation, seem to be associated with greater support from backers. If properly managed, this environmental driver might be sustained over time, resulting in long-lasting relationships with mindful customers.

6.5 Methodology

6.5.1 Data Collection

The dataset was built by selecting the fashion category in Kickstarter. This search yielded 38,217 projects which were filtered by their condition, only successful projects, and the pledge amount. Only successful projects with a pledge amount of more than $100,000 were selected so that not only could the projects selected allow a good alignment with the purpose of this study, but also a large sample size was guaranteed (Hong et al., 2021). As a result, a sample of 341 projects was analyzed.

As for the coding process, the following information was extracted for each project: title, description, creators, number of backers, pledge amount, goal amount, location, category, most successful segment per number of backers, pledge amount of the most successful segment, and funding period. Fieldwork took place during September 2023.

Out of this information, some categories were built to facilitate a better interpretation of the results. These new categories were: gender of creators, main keywords, and growth rate. The methodology chosen to study this relationship was thematic content analysis, which is discussed next.

6.5.2 *Thematic Content Analysis (TCA)*

TCA is the most foundational of qualitative analytic procedures (Anderson, 2007; Green & Thorogood, 2005). An inductive approach was favoured to start with: this helped with theme development to then perform a deductive approach. Thus, the whole database ($N = 341$ projects) was further subdivided into categories using the data column "category". Four categories were identified: *accessories, underwear, apparel*, and *footwear*. These four categories comprised almost 90% of the whole database. Forty (40) projects were dismissed due to their miscellaneous nature and difficulty in categorization (e.g., a nonprofit dedicated to building a sustainable fashion industry). The weight of the categories' contributions is presented in Table 6.1.

As suggested by previous work (Defazio et al., 2021), content analysis was carried out on the field "descriptions", indicating that understandable and relatable descriptions correlated with attraction of funding, particularly in social causes. Typically, these descriptions contain enough words to make coding and thematic extraction possible. Thematic content analysis was performed using descriptions for each category separately. Given the decision to use categories, the words naming the type of product in the column "type of clothing" were deleted from the descriptions to get a

Table 6.1 Projects classified by category

Category	Number of projects in category	% of projects versus total database (%)
Accessories	91	26.7
Underwear	19	5.6
Apparel	139	40.8
Footwear	52	15.2
Other	20	5.9
Unspecified	20	5.9
Total	**341**	**100**

Source authors

better sense of codes. In this manner, names such as "bags" and "wallets" (predominant in *accessories*); "underwear" (already classified as such); "hoodies", "pants", and "shirts" (predominant in *apparel*) and "shoes" and "boots" (predominant in *footwear*) were erased from the database prior to thematic coding.

6.6 Results

A comparative analysis of initial themes using an inductive approach is outlined in Table 6.2. There are striking similarities in how descriptions are collectively used to reflect sustainable behaviour, emphasizing functional benefits as a measure of quality and presence of sustainable materials in all four categories. All categories make it clear that functional benefits are adapted to the relevant category–for example, grip in boots, or odour control in *underwear*. However, the sense of functional benefits transcends the categories in which they appear and evolve into functional benefits that symbolize quality, a core demand for sustainable products (Aakko & Niinimäki, 2022). Another similarity that is important for sustainable consumption and which appears in two of the four categories, namely, *accessories* and *footwear*, is the presence of social benefits (Sinha et al., 2022) in the form of empowerment of local communities, heritage and history and a mention of origin, most typically as a country. In both, we also see specific sustainability buzzwords (in particular "eco-friendly").

The most striking differences refer to lifestyle. Whilst design and style are important for the *accessories* category, *apparel* shows a clear predominance of text regarding adventure, exploration, and travel, which in turn explains the overuse of functional benefits such as "comfort" and "fit". The same strong emphasis on functional benefits is shown in the *underwear* category. However, emotional benefits are distinctive in the *footwear* category, which indicates a wide range of lifestyle formulations per category, from more objective cues (e.g., comfort) to subjective propositions where emotional engagement might be pursued.

When matched against the theoretical framework provided by Heidenreich et al. (2022), the similarity of these inductive themes shown with the theory allowed a deductive approach. Thus, common themes summarized in Table 6.2 can be distinguished in four areas: (1) material sources and their use from pure natural to high-tech processed; (2) functional benefits–mostly adapted to the category–and emotional benefits derived from them; (3) benefits of alternative modes of consumption, for example, empowerment of local communities; and (4) enhancement of quality. All these themes are represented in the literature; the first three are described by Heidenreich et al. (2022). Quality enhancement as a priority beyond success has also been previously identified (Dalla Chiesa et al., 2022).

A deep dive into the three themes described by Heidenreich et al. (2022) yields interesting results. From the use of materials, summarized in Table 6.3, it appears that each category interprets sustainability in different ways, though all four categories use natural sustainable materials. *Apparel* focuses on recyclable materials

Table 6.2 Comparative themes across categories (predominant themes first)

Accessories	Underwear	Apparel	Footwear
Strong emphasis on design and style	Strongest emphasis on functional delivery (benefits) around water, sweat e.g. quick drying, odour-killing	Relevance for lifestyle–adventure, exploration, travelling	Nice combination of craftsmanship and local communities
Handcrafted products	Secondary functional benefits in comfort and protection	Technical performance oriented	Sense of time ever present–legacy, lifetime, generations. Bridging the past and the future
Sustainable materials (leather, enamel, ceramic)	Clear sense of emotional benefits around sensations	Functional benefits around fit, particularly comfort	Quality enhancement throughout: durability, sturdiness, grip
New durable materials (sapphire, titanium)	Sustainable materials around technology (e.g. nanotech, underlux) and recyclable (underlux)	Clear mention of features related to staying dry when outdoors	Functional benefits around comfort and adaptability
Emotional benefits around genuineness	Sustainable materials around vegetables (eucalyptus, peppermint)	Sustainable materials: wool, leather; sustainable processes; selvedge	Other functional benefits around high tech–breathable, wind and waterproof, anti-bacterial
Functional benefits around durability and affordability	Clear mention of sustainability benefits (e.g. eco-friendly)	Smart technology: RFID	Specific sustainability messages around eco-friendliness (ocean, recycling, forests)
	Sense of innovation (new-generation, life-changing)	Pressure to ensure quality: warranty, certified, guarantee	Emphasis on locally made with specific mention of where (e.g. Italy, Kenya, Brazil, etc.)
		Recycled materials, particularly plastic	Non-toxic sustainable materials (vegan, hemp, sugarcane) design and style

Source authors

(enamel, ceramic) or eco-friendly materials (sapphire, titanium). *Underwear* focuses on two types of sustainable materials: either pure natural (e.g. eucalyptus, beech trees) or high-tech sustainable materials that are process-related, the other dimension identified. In this way, bioceramic is obtained from wastewater, so it is not only economical but sustainable to manufacture; silver tech is a sustainable, low-concentration, recycled silver chloride treatment and nanotechnology enhances environmental sustainability processes by making more efficient use of rare and strategic materials, substituting hazardous materials (Ciambelli et al., 2020).

Apparel uses selvedge processes for denim and jeans, involving weaving in the old way, but is one of the four categories, overall, least invested in sustainable materials and also the only category in which smart technology appears. *Footwear* is heavily invested in natural sustainable materials, mostly plant based and thus vegan. The analysis of materials also helps to identify and classify the processes used, discerning whether they in themselves are sustainability focused, and an early indicator of mindful production (Finch, 2010). The only two categories that show sustainable processes in place are *underwear* and *apparel*–of the two, *apparel* has some focus on recycled plastic bottles, but it is *underwear* that clearly leads crowdfunding efforts for sustainable production processes.

As per the second theme, the analysis of functional and emotional benefits also yields interesting differences between the four categories under study (see Table 6.4).

Table 6.3 Material sources across categories

Accessories	Underwear	Apparel	Footwear
Natural sustainable	Natural sustainable	Natural sustainable	Natural sustainable
Leather	*Eucalyptus*	*Wool*	*Vegan*
Wood	*Peppermint*	*Selvedge**	*Organic*
	Beech trees		*Hemp*
Recyclable materials		Recycled Materials	*Sugarcane*
Enamel	High-Tech Sustainable*	*Plastic Bottles**	*Wool*
Ceramic	*Bioceramic*		*Leather*
	Silver tech	Smart Technology	
Eco-friendly materials	*Nanotech*	*RFID*	High-Tech Sustainable*
Sapphire	*Thermal*		*Microfibre*
Titanium	*Micro-modal*		*Breathable*
	Other High-Tech		Other High-Tech
	Underlux		*Quick-dry*
	3D Thermal		
	Other materials	Other materials	
	Silk	*Silk*	

Source authors

* The use of * refers to sustainable processes behind the materials

Functional benefits are category-led; this is particularly obvious in the case of *underwear* (e.g., anti-odour, absorbent, breathable) and *footwear* (e.g., durability, sturdiness, waterproof). In *apparel*, the benefits highlighted are consistent with the dominant theme of adventure and outdoor life (e.g., moisture-check, submersible, waterproof). The predominance of watches in *accessories* explains the presence of benefits such as automatic or mechanical.

Emotional benefits highlight comfort as a primary motivation and style as secondary; then each category ladders up to category-specific benefits in such a way that softness, protection, and support are important for *underwear* vs. versatility, adaptability and grip which resound for *footwear*. There is one exception, however, the *accessories* category, where style is predominantly emphasized in what could be projected as a search for uniqueness.

Table 6.4 Functional, emotional and social benefits

Accessories	Underwear	Apparel	Footwear
Functional benefits			
Durable	Absorbent	Technical	Fold and not crease
Affordable	Odour-killing, anti-odour	Wrinkle-free	Durable, extreme
Lightweight	Breathable	High-performance	Sturdy
Waterproof	Quick-dry	Moisture-check	High-tech
Mechanical	Period-proof	Odour-killing	Anti-bacterial
Automatic		Stain-avoidance	Wind-proof
		Submersible	Hypoallergenic
		Waterproof	Waterproof
		Repelling	
Emotional benefits			
Stylish	Defy gravity	Stretches	Stylish, sophisticated
Genuine	Sensation, feel	Comfortable	Customizable
Vintage	Softness	Softness	Versatile
Awesomeness	Protection		Strong, grip
Minimalistic	Supportive, posture		Adaptable, moldable
	Slimming		Comfortable
	Comfortable		Buyback
Social benefits			
Handmade	Not mentioned	Not mentioned	Locally-made
Handcrafted			Legacy
Embroided			Helping communities
Story			Focus on world
			Past, present, tomorrow

Source authors

Perhaps the most interesting finding, however, is the highlight of social benefits for *accessories* and *footwear* but not the other two categories. This is consistent with theme extraction, for it is in these two categories that specific mention is made regarding helping local communities. In *accessories*, this is "light" and mostly related to crafting of sorts, including embroidery. The extent to which these social benefits are formulated is also worth mentioning. From a handmade locally-produced accessory to a focus on a globally-produced type of footwear.

Two of the categories studied also explained some benefits of alternative modes of consumption, the third theme analysed, in terms of social benefits, mainly regarding improvement of local communities. *Accessories* have a whole theme around embroidered, handmade and handcrafted products that improve situations of local communities. In *footwear*, this is a clear theme that involves three threads: (1) helping communities; (2) bridging the past with the present across generations, a sense of timelessness; and (3) upgrading the past to ensure a better present and future for a better world. It is by highlighting these social benefits that the value proposition becomes fully sustainable, or more related to the concept of the circular economy and the social value of sustainable consumption.

The application of the Morseletto framework (Morseletto, 2020) and its ten circular economy strategies helps to complement the analysis, particularly the extent to which these initiatives are real targets for a circular economy. Thus, our results show that only *apparel* is fully focused on keeping within these targets, with 89% of the projects classified under one of the Morseletto's Rs (see Table 6.5). The other categories show weaker targets, with only one out of every two projects fulfilling the criteria; this is particularly "poor" in the case of *accessories*.

Table 6.5 Application of Morseletto's framework for a circular economy (2020)

	Accessories	Underwear	Apparel	Footwear
Target: smarter product use and manufacture				
Refuse			2%	
Rethink	50%	40%	18%	37%
Reduce	9%	50%	77%	22%
Target: extend lifespan of product and its parts				
Reuse				
Repair				4%
Refurbish	26%	10%		19%
Remanufacture	9%		4%	
Repurpose	3%			
Target: useful application of materials				
Recycle	3%			19%
Recovery				
Proportion of Projects Classified	37%	50%	89%	52%

Source authors

All categories peak at "Smarter Product Use and Manufacture" (Refuse, Rethink and Reuse), with over 50% of all initiatives classified under that area. This is an important finding, particularly given that these targets take place when products are conceived, designed and developed and for that reason enable and aid the transformation towards a circular economy as a circular rather than linear process (Morseletto, 2023).

The close connection of this target to design appears in the *apparel* category, which primarily focuses on making sure all projects are designed with circular economy targets in mind. Since 77% of *apparel*'s projects are classified under Reduce, this category designs clothing (pants, denim, shirts etc.) specifically to use fewer natural resources. This is very evident in the case of denim, for instance, with a strong presence of selvedge as a process. *Underwear* follows a very similar pattern, with 90% of all projects classified under the "Smarter Product Use and Manufacture" target; this makes sense considering the materials used and the emphasis on high-tech sustainable processes.

Accessories and *footwear* provide an alternative reading. Despite showing a bias towards smarter product use, both of these categories focus on the extension of the lifespan of the product, reducing the intent to repurchase. Both categories also show a predominance of Refurbish, which emphasizes the objective of restoring old products and bringing them up to date, typically through replacing parts, increasing the value of the product (and thus its quality). This is the objective of all pins products in the *accessories* category but also a large number of boots that are inspired by past designs and past production processes that see the light in a different way. *Footwear* has an added presence in Recycle, using materials extracted from discarded products to create new options: this is particularly relevant in the case of plastic bottles.

A second analysis was carried out separating each category by project using the compliance with one of Morselettos' Rs giving as a result two subsets, R classified and not classified. Interesting differences were found in the thematic content for all categories, particularly *underwear* and *footwear* depending on whether or not the project had previously been classified as "R Classified". The case of *underwear* (Table 6.6) is particularly interesting because it shows how sustainability-led initiatives were very clear on functional benefits regarding quality but also on the high-tech processes previously mentioned e.g., nanotech, silver tech etc.) and the natural materials used (eucalyptus, peppermint etc.). Almost all product types in these classified groups included bottom *underwear* (pants, briefs) and were relevant for males and females alike. However, the projects that were not possible to classify using Morseletto showed primarily female top *underwear* (bras, bustiers etc.) and thus focused on the functional benefits required for that category that falls under adequate support, defy gravity, shape and posture and perkiness of breasts. As a result, emotional benefits regarding sensations (floating, comfort) were stressed.

A similar output was produced for *footwear*. Huge differences appeared between the two subsets, as discussed in Table 6.7. The clearest difference between the two subsets lies in materials and processes used, particularly in the presence of leather, which is only present in projects classified under one of the "Rs", but also of most

Table 6.6 Underwear–subclassification of themes by type of project

Overall Themes	R classified	R not-classified
360° Sustainability-focus		
Strongest emphasis on functional delivery (benefits) around water, sweat e.g. quick dry, odour-killing	Functional Benefits as described in Table 6.4	Only functional benefit specified around period-proof and menstruation
Secondary functional benefits in comfort and protection		Strong emphasis on functional benefits around support, posture, shape
Clear sense of emotional benefits around sensations		Strong mention of emotional benefits around sensations (feel, floating etc.)
Sustainable materials around technology (e.g. nanotech, underlux) and recyclable (underlux)	High-Tech (process-related)	
Sustainable materials around vegetable (eucalyptus, peppermint)	Natural sustainable materials	
Clear mention of sustainability benefits (e.g. eco-friendly)	Specific mention of sustainable benefits (eco-friendly)	
Sense of innovation (new-generation, life-changing)		

Source authors

of the natural sustainable materials used (hemp, sugarcane, plastic bottles etc.). R-classified projects are clearly eco-friendly and use a wide range of words to transmit this commitment to sustainability. The most important difference, however, lies in the emphasis of R-classified projects on the social benefits of alternative forms of consumption, both in terms of craftsmanship and handmade footwear to strengthen and empower local communities but also in terms of legacy and bridging generations, updating the past to bring to the present something that is more sustainable and up to date. On the other hand, non-classified projects focus almost exclusively on beauty, design and style; but also on quality around functional benefits of wind and waterproof, comfort and grip.

6.7 Discussion

This chapter set out to examine the extent to which entrepreneurs are spearheading the promotion of sustainable modes of consumption, captured under the concept of mindful consumption, through a detailed analysis of textile-led projects in Kickstarter, a reward crowdfunding platform. Our analysis shows that evidence for a sustainability-focused mindset is consistent across all textile categories examined:

Table 6.7 Footwear–subclassification of themes by type of project

Overall themes	R-classified	R-not-classified
These boots were made for walking…		
Nice combination of craftsmanship and local communities	Handcrafted, handmade, crafted local communities strengthen communities empowering local communities	
Sense of time ever present–legacy, lifetime, generations. Bridging the past and the future	Upgraded founded in 1907, now brought back to life original bushman continue the legacy Built on generations	
Quality enhancement throughout: durability, sturdiness, grip	Crafted to high quality designed for persistence	Top quality premium quality Eva foam* (non-sustainable) Extreme grip
Functional benefits around comfort and adaptability	Comfort	Comfort, durability
Other functional benefits around high tech–breathable, wind and waterproof, anti-bacterial	Breathable, anti-bacterial	Waterproof custom-heat moldable to form to your feet windproof
Specific sustainability messages around eco-friendliness (ocean, recycling, forests)	Eco-friendly upcycled sourced-locally buyback Restoring forests Recycled materials made of plastic bottles	
Emphasis on locally made with specific mention of where (e.g. Italy, Kenya, Brazil, etc.)	Brazil, Boston, Italy, Kenya 100% sourced	New textiles from Guatemala
Natural sustainable materials (vegan, hemp, sugarcane)	Leather is huge, then followed by other vegetable natural sustainable materials Sugarcane, hemp, vegan Wool	No mention of leather
Style and design		Timelessness style classic design sophisticated colourful fashion-look minimalist

Source authors

accessories, *underwear*, *apparel*, and *footwear*. This mindset is made evident through four important topics extracted from the theoretical framework provided by Heidenreich et al. (2022): materials and processes used, functional and emotional benefits communicated, and social benefits of such alternative modes of consumption. The concept of quality is implicit in the constant mention of functional benefits which are also points of parity for each of the categories analysed. Overall, functional and emotional benefits seem to be the "must-have" elements to be considered sustainable quality, whilst social benefits are "nice-to-have" and add to the quality proposition, a combination of benefits that should not be neglected when fostering alternative modes of consumption that are compatible with the triple bottom line of businesses (Elkington, 2018).

Despite this consistency, the evidence for a circular economy is irregular, at best. The classification of projects using the "Rs" proposed by Morseletto (2020) shows preliminary qualitative evidence that not all categories are equally invested in the promotion of mindful consumption to start with. The relationship between the circular economy and Morseletto's Rs (Morseletto, 2020) and Mindful Consumption (Sheth et al., 2011) for each category is summarised in Table 6.8.

The high proportion of projects in *apparel* that can be classified under one of the Rs is a bit of a red herring, since there is no evidence in the analysis of text that there is an investment in either sustainable materials or processes beyond wool and selvedge. Indeed, almost 90% of the projects classified as *apparel* under Morseletto's "Rs" are in Rethink and Reduce, suggesting the target of reaching a smarter product use and manufacture as the only target pursued for an integral circular economy strategy. This would suggest the type of mindful consumption for *apparel* is acquisitive in nature, aimed at reducing consumption rather than rethinking consumption. A similar conclusion is reached when looking at *accessories*, where the small proportion of sustainability-led projects and the emphasis on style and design suggest an aspirational mindful consumption.

It is in the other two categories, *underwear* and *footwear*, that more substantial evidence for mindful consumption is found, in the form of repetitive mindful consumption. In both these categories, there is a neat combination of natural sustainable materials and sustainable processes to match, particularly in the case of *underwear*. However, it is *footwear* that becomes the most interesting category to study,

Table 6.8 Classification of categories, circular economy and mindful consumption

	Accessories	Underwear	Apparel	Footwear
Caption for category	*Handcrafted stylish products*	*360º Sustainability focus*	*Reduce but not abuse*	*These boots were made for walking …*
Morseletto's Rs	Rethink, Reduce, Refurbish, Remanufacture, Repurpose, Recycle	Rethink, Reduce, Refurbish	Refuse, Rethink, Reduce, Remanufacture,	Refuse, Rethink,, Repair, Refurbish, Recycle
Primary Circular Economy Group	Smarter product use & manufacture	Smarter product use & manufacture	Smarter product use & manufacture	Smarter product use & manufacture
Mindful Mindset	Self community	Self environment	Self	Self community environment
Mindful Consumption Behaviour	Aspirational	Repetitive	Acquisitive	Repetitive

Source authors

since it is the only category where strong evidence for social benefits of alternative modes of consumption is found, mostly around the enhancement and empowerment of local communities but also in the bridge between the past and the future: bringing old, handcrafted boots (primarily) and shoes up to date, which explains why the category peaks at Refurbish. It is also the only category in which there is specific mention of Recycle behaviour in the use of plastic to create new *footwear*. The contrast with non-classified projects provides additional evidence that, when designed with a sustainable mindset, this category emphasizes the right combination of elements beyond more fashion-led statements regarding style and design, specifically Italian design.

Taking the leap from circular economy evidence, mainly in the form of smarter product use and manufacture to mindful consumption, yields interesting comparisons. Mindful consumption as described by Sheth et al. (2011) takes into account two different elements: mindset and behaviour. All four categories show a mindful mindset focused on self or self + community (*accessories*) or self + environment (*underwear*). Only one category, *footwear*, focuses on self + community + environment.

From this analysis we can discuss a possible roadmap for the promotion of mindful consumption, the objective of this chapter. Mindful consumption requires a combination of the following elements: materials used, which have to be natural sustainable materials *plus* method of production, which needs to be sustainability-focused *plus* the communication of functional materials, which yield credibility as to the quality of the proposal *plus* benefits beyond consumption, namely social benefits.

Thanks to the combination of these four elements: (1) materials, (2) production methods, (3) communication and (4) functional, emotional and social benefits, a mindful mindset that cares about self + community + environment is ensured and matched by a mindful consumption behaviour and a mindful production and process. Of the categories examined, only one, *footwear*, fully meets all the criteria.

Despite this match, it is *underwear* which clearly sets the tone for mindful consumption in the textile industry. In the future, this category could benefit from highlighting its role in social benefits of alternative sources of consumption for a full impact. Less importance should be given to design and beauty, and more emphasis should be placed on the impact of the use of natural sustainable materials and high-tech sustainable processes on the communities served.

6.8 Theoretical Implications

Regarding the analysis of the category that might fully represent a mindful consumption mindset, *footwear*, a main theoretical contribution is offered. For new business ideas to fully comprise the focus on self + community + environment, mindful modes of consumption should be promoted from its roots, that is, from a "*mindful production*" mindset and implementing a "*mindful process*". This is aligned with the barriers encountered when facing the transition to the circular economy from the lens of an integral process rather than a linear set of outcomes (de Aguiar Hugo et al., 2021; Morseletto, 2023). Moreover, in the discussion of barriers and opportunities for this transition, awareness is seen as one of the enablers to foster a new paradigm of consumption, a more sustainable mindful consumption.

In addition, this study assesses the state a mindful consumption mindset and behaviour. Based on this study, efficiency in the use and manufacture of products is emphasized compared with other strategies. This evidence points out that, even in the context of social innovation platforms, an integral approach to circular economy strategies is still lacking. Only when a balance between usage and manufacturing, design and resources extraction is achieved, could alternative modes of consumption and production be guaranteed. This highlights the lack of knowledge, tools and methodologies needed to achieve a more balanced approach to circular economy strategies.

6.9 Managerial Implications

The thematic analysis also offers practical implications for entrepreneurs in the design of sustainable products and services. Given that previous research evidenced that campaign descriptions positively correlate with the attraction of funding to social causes (Defazio et al., 2021), the detailed analysis performed in this study offers clear and distinctive sustainability cues facilitating full alignment between a responsible business and its social capital, particularly its backers.

This study emphasizes that functional benefits which are aligned with its emotional benefits seem to represent a measure of quality while the presence of social benefits represents communities' empowerment as a lifestyle manifestation. Moreover, our results also suggest that social and emotional benefits are prioritized versus design and style. This is also backed up by the R-non-classified projects where beauty and design seem insufficient to constitute the basis of a consistent sustainable proposition. This also supports the notion of how building identity around uniqueness and authenticity should be closely linked to alternative modes of consumption and production.

Brands are encouraged to leverage these findings to strengthen their internal and external communications and marketing messages, emphasizing purpose and values as measures of their authenticity. Overall, the study emphasizes the importance of conveying a balanced combination of functional, emotional, and social benefits, along with clear information about materials, production, and processes to foster a lifestyle shared between the self, the community, the environment, and the business. By avoiding sustainability buzzwords and maintaining coherence between actions and messaging, companies can reinforce customer engagement in the short term and over time. Overall, the managerial implications outlined in this chapter aim to offer valuable guidance for entrepreneurship and innovation in promoting mindful consumption and production.

6.10 Concluding Remarks

The most striking conclusion that can be extracted out of this analysis is the lack of a *mindful production* mindset in the ideation of sustainable business ideas. Only one category of the one analysed seems to clearly follows this trend, that is, *footwear.* This idea is supported when innovative business ideas, such as those at Kickstarter, are confronted with the Morseletto framework (Morseletto, 2020). Only one category, *apparel*, seems to transcend from usage, which could be more related to how mindful consumption is conceived, to design and raw materials of resources. This fact suggests that a mindful mindset needs to be implemented in all stages to be able to capture the extent of how a business idea is first conceived. The exception of the *underwear* category should be taken as a reference. This is a perspective that might require the exploration of other forms of mindful mindsets and, in this, mindful production seems a line of research worth pursuing further.

References

Aakko, M., & Niinimäki, K. (2022). Quality matters: Reviewing the connections between perceived quality and clothing use time. *Journal of Fashion Marketing and Management: An International Journal, 26*(1), 107–125. https://doi.org/10.1108/JFMM-09-2020-0192

Aleem, A., Loureiro, S. M. C., & Bilro, R. G. (2022). Luxury fashion consumption: a review, synthesis and research agenda. *Spanish Journal of Marketing-ESIC,* (ahead-of-print).

Anderson, R. (2007). Thematic content analysis (TCA). *Descriptive Presentation of Qualitative Data, 3*, 1–4.

Areias, S., Disterheft, A., & Gouveia, J. P. (2023). The role of connectedness in pro-environmental consumption of fashionable commodities. *Sustainability, 15*(2), 1199. https://doi.org/10.3390/su15021199

Armstrong, C. M. J. (2021). Fashion and the Buddha: What Buddhist economics and mindfulness have to offer sustainable consumption. *Clothing and Textiles Research Journal, 39*(2), 91–105. https://doi.org/10.1177/0887302X20917457

Belleflamme, P., Lambert, T., & Schwienbacher, A. (2014). Crowdfunding: Tapping the right crowd. *Journal of Business Venturing, 29*(5), 585–609. https://doi.org/10.1016/j.jbusvent.2013.07.003

Böckel, A., Hörisch, J., & Tenner, I. (2021). A systematic literature review of crowdfunding and sustainability: Highlighting what really matters. *Management Review Quarterly, 71*, 433–453. https://doi.org/10.1007/s11301-020-00189-3

Cai, Y. J., & Choi, T. M. (2020). A United Nations' sustainable development goals perspective for sustainable textile and apparel supply chain management. *Transportation Research Part E: Logistics and Transportation Review, 141*, 102010. https://doi.org/10.1016/j.tre.2020.102010

Ciambelli, P., La Guardia, G., & Vitale, L. (2020). Nanotechnology for green materials and processes. In *Studies in surface science and catalysis* (Vol. 179, pp. 97–116). Elsevier. https://doi.org/10.1016/B978-0-444-64337-7.00007-0

Cricelli, L., Grimaldi, M., & Vermicelli, S. (2022). Crowdsourcing and open innovation: A systematic literature review, an integrated framework and a research agenda. *Review of Managerial Science, 16*(5), 1269–1310. https://doi.org/10.1007/s11846-021-00482-9

Dalla Chiesa, C., Pavlova, A., Lavanga, M., & Pysana, N. (2022). When fashion meets crowdfunding: exploring sustainable and innovative features of online campaigns. *Journal of Fashion*

Marketing and Management: An International Journal. https://doi.org/10.1108/JFMM-03-2021-0068

de Aguiar Hugo, A., de Nadae, J., & da Silva Lima, R. (2021). Can fashion be circular? A literature review on circular economy barriers, drivers, and practices in the fashion industry's productive chain. *Sustainability, 13*(21), 12246. https://doi.org/10.3390/su132112246

Defazio, D., Franzoni, C., & Rossi-Lamastra, C. (2021). How pro-social framing affects the success of crowdfunding projects: The role of emphasis and information crowdedness. *Journal of Business Ethics, 171*, 357–378. https://doi.org/10.1007/s10551-020-04428-1

Del Pilar Quiroz Galvan, M., Fritz, M. M., Šimunović, N., Stern, T., & Rauter, R. (2021). Overcoming sustainability challenges with non-profit organisations? Insights from the apparel supply chain. *Supply Chain Forum: An International Journal, 22*(2), 115–135. (Taylor & Francis, Apr 2021). https://doi.org/10.1080/16258312.2021.1911568

Del Prete, M. (2022). Mindful sustainable consumption and sustainability chatbots in fast fashion retailing during and after the COVID-19 pandemic. *Journal of Management & Sustainability, 12*, 19. https://doi.org/10.5539/jms.v12n1p19

Deng, L., Ye, Q., Xu, D., Sun, W., & Jiang, G. (2022). A literature review and integrated framework for the determinants of crowdfunding success. *Financial Innovation, 8*(1), 41. https://doi.org/10.1186/s40854-022-00345-6

Di Vaio, A., Hassan, R., D'Amore, G., & Tiscini, R. (2022). Responsible innovation and ethical corporate behavior in the Asian fashion industry: A systematic literature review and avenues ahead. *Asia Pacific Journal of Management*, 1–45. https://doi.org/10.1007/s10490-022-09844-7

Dzhengiz, T., Haukkala, T., & Sahimaa, O. (2023). (Un) Sustainable transitions towards fast and ultra-fast fashion. *Fashion and Textiles, 10*(1), 19. https://doi.org/10.1186/s40691-023-00337-9

Elkington, J. (2018). 25 years ago I coined the phrase "triple bottom line." Here's why it's time to rethink it. *Harvard Business Review, 25*, 2–5.

European Parliament. (2023). *EU Strategy for Sustainable and Circular Textiles.* https://www.europarl.europa.eu/doceo/document/TA-9-2023-0215_EN.html

Fan, K. K., & Chang, Y. T. (2023). Exploring the key elements of sustainable design from a social responsibility perspective: A case study of fast fashion consumers' evaluation of green projects. *Sustainability, 15*(2), 995. https://doi.org/10.3390/su15020995

Finch, M. L. (2010). Food, taste, and American religions. *Religion Compass, 4*(1), 39–50. https://doi.org/10.1111/j.1749-8171.2009.00166.x

Green, J., & Thorogood, N. (2005). Analysing qualitative data. *Principles of Social Research*, 75–89.

Gupta, S., & Verma, H. (2019). Mindful consumption behaviour: Scale development and validation. *Asian Journal of Multidimensional Research (AJMR), 8*(5), 271–278. https://doi.org/10.5958/2278-4853.2019.00200.3

Heidenreich, S., Freisinger, E., & Landau, C. (2022). The dark side of business model innovation: An empirical investigation into the evolvement of customer resistance and the effectiveness of potential countermeasures. *Journal of Product Innovation Management, 39*(6), 824–846. https://doi.org/10.1111/jpim.12627

Hong, L. J., Huang, Z., & Lam, H. (2021). Learning-based robust optimization: Procedures and statistical guarantees. *Management Science, 67*(6), 3447–3467. https://doi.org/10.1287/mnsc.2020.3640

Jung, Y. J., & Kim, Y. (2023). Research trends of sustainability and marketing research, 2010–2020: Topic modeling analysis. *Heliyon, 9*(3). https://doi.org/10.1016/j.heliyon.2023.e14208

Lipusch, N., Dellermann, D., Bretschneider, U., Ebel, P., & Leimeister, J. M. (2020). Designing for crowdfunding co-creation: How to leverage the potential of backers for product development. *Business & Information Systems Engineering, 62*, 483–499. https://doi.org/10.1007/s12599-019-00628-w

Mahmud, S. H., Anuar, M. M., Halim, M. A. S. A., & Yaakop, A. Y. (2019). The influence of self care on mindful consumption behaviour. *The Journal of Social Sciences Research, 5*(1), 94–100. https://doi.org/10.32861/jssr.51.94.100

Mollick, E. (2014). The dynamics of crowdfunding: An exploratory study. *Journal of Business Venturing, 29*(1), 1–16. https://doi.org/10.1016/j.jbusvent.2013.06.005

Moon, C. J. (2018). Contributions to the SDGs through social and eco entrepreneurship: New mindsets for sustainable solutions. *Entrepreneurship and the Sustainable Development Goals*, 47–68. https://doi.org/10.1108/S2040-724620180000008008

Mora-Cruz, A., & Palos-Sanchez, P. R. (2023). Crowdfunding platforms: a systematic literature review and a bibliometric analysis. *International Entrepreneurship and Management Journal*, 1–32. https://doi.org/10.1007/s11365-023-00856-3

Morseletto, P. (2020). Targets for a circular economy. *Resources, Conservation and Recycling, 153*, 104553. https://doi.org/10.1016/j.resconrec.2019.104553

Morseletto, P. (2023). Sometimes linear, sometimes circular: States of the economy and transitions to the future. *Journal of Cleaner Production*, 136138. https://doi.org/10.1016/j.jclepro.2023.136138

Nilsson, F., & Göransson, M. (2021). Critical factors for the realization of sustainable supply chain innovations-model development based on a systematic literature review. *Journal of Cleaner Production, 296*, 126471. https://doi.org/10.1016/j.jclepro.2021.126471

Niinimäki, K., Peters, G., Dahlbo, H., Perry, P., Rissanen, T., & Gwilt, A. (2020). The environmental price of fast fashion. *Nature Reviews Earth & Environment, 1*(4), 189–200. https://doi.org/10.1038/s43017-020-0039-9

Ostermann, C. M., da Silva Nascimento, L., Steinbruch, F. K., & Callegaro-de-Menezes, D. (2021). Drivers to implement the circular economy in born-sustainable business models: A case study in the fashion industry. *Revista De Gestão, 28*(3), 223–240. https://doi.org/10.1108/REGE-03-2020-0017

Papasolomou, I., Melanthiou, Y., & Tsamouridis, A. (2023). The fast fashion vs environment debate: Consumers' level of awareness, feelings, and behaviour towards sustainability within the fast-fashion sector. *Journal of Marketing Communications, 29*(2), 191–209. https://doi.org/10.1080/13527266.2022.2154059

Pero, M., Arrigo, E., & Fionda-Douglas, A. (2020). Sustainability in fashion brands. *Sustainability, 12*(14), 5843. https://doi.org/10.3390/su12145843

Ray, S., & Nayak, L. (2023). Marketing sustainable fashion: Trends and future directions. *Sustainability, 15*(7), 6202. https://doi.org/10.3390/su15076202

Sheth, J. N., Sethia, N. K., & Srinivas, S. (2011). Mindful consumption: A customer-centric approach to sustainability. *Journal of the Academy of Marketing Science, 39*(1), 21–39. https://doi.org/10.1007/s11747-010-0216-3

Sinha, P., Sharma, M., & Agrawal, R. (2022). A systematic review and future research agenda for sustainable fashion in the apparel industry. *Benchmarking: An International Journal*, (ahead-of-print). https://doi.org/10.1108/BIJ-02-2022-0142

Wren, B. (2022). Sustainable supply chain management in the fast fashion Industry: A comparative study of current efforts and best practices to address the climate crisis. *Cleaner Logistics and Supply Chain, 4*, 100032. https://doi.org/10.1016/j.clscn.2022.100032

Zhang, X. Y., Tu, J. C., Gu, S., Lu, T. H., & Yi, M. (2022). Construct and priority ranking of factors affecting crowdfunding for green products. *Processes, 10*(3), 480. https://doi.org/10.3390/pr10030480

Chapter 7
Strategies in the Luxury Fashion Industry Integrating the Circular Economy to Impact SDG12 and SDG13

Belén López-Vázquez and Celia Rangel-Pérez

7.1 Introduction

Sustainability has become an increasing strategy for companies worldwide because of the global risks for the planet. In this context, sustainable business models are important in large luxury companies, especially the fashion luxury conglomerates (Campos et al., 2020; Ozdamar-Ertekin, 2019), as leaders in the global market that also promote the fight against climate change (Osburg et al., 2021).

Moreover, sustainability means that companies are implementing practices in the Corporate Social Responsibility (CSR) strategy to contribute to a better society by connecting the corporate culture, perceptions, and expectations of their stakeholders (Pedersen et al., 2018). However, today luxury conglomerates need to show the impact of their actions (Rukhaya et al., 2021) to reduce CO2 emissions and protect the environment, leading this process in the market (Jia et al., 2020) and following regulations in the European Union, such as the Green Deal and the 2030 Agenda promoted by United Nations with the 17 Sustainable Development Goals and their 169 targets.

Consequently, luxury firms innovate using a circular economy in their processes implementing new sustainable solutions where production is based on regeneration and the use of different recycled materials for new products, such as recycling materials (Campos et al., 2020; Thorisdottir & Johannsdottir, 2020; Willersdorf et al.,

B. López-Vázquez (✉)
ESIC University, Pozuelo de Alarcon, Spain
e-mail: belen.lopez@esic.university

ESIC, Business & Marketing School, Madrid, Spain

C. Rangel-Pérez
Universidad Complutense de Madrid, Madrid, Spain
e-mail: cerangel@ucm.es

J. F. Gallego-Nicholls and A. Carrilero-Castillo (eds.), *Textile Industry and the SDGs*, SDGs and Textiles, https://doi.org/10.1007/978-981-97-5073-3_7

2020) and impacting consumers' perceptions (Kunz et al., 2020) who are demanding sustainable practices in luxury brands (Pencarelli et al., 2020).

Furthermore, it is essential in this process to explore how luxury firms use communication to show their responsibility in all the areas of their strategy to influence stakeholders' perceptions and to interact with them, especially in the current digital context (San Miguel et al., 2021). On one side, the communication strategy can increase the visibility of a company's sustainable activities (Da Giau et al., 2016) and improve the credibility of the company based on its environmental and social value (San Miguel et al., 2021). Additionally, some scholars mention a risk based on cosmetic communication to improve the perception of the firm, luxury firms are increasing their presence through traditional and online platforms explaining their sustainable business and activities (Abitbol et al., 2019; Mann et al., 2014; Siano et al., 2016; Troise & Camilleri, 2021).

However, companies from all sectors include in their annual CSR/Sustainability reports increasingly extensive information on good practices in sustainability and their impact on each of the 17 SDGs and their 169 goals with concrete actions and the global impact of the company, as the sum of all those actions. Consequently, the communication strategy should be aligned with the CSR strategy of companies, in particular luxury firms, with the objective of showing how these firms are innovating from a sustainable perspective.

This chapter aims to study how LVMH is strategically communicating its sustainability efforts, specifically those related to the circular economy, SDG 12 (Responsible consumption and production) and SDG 13 (Climate action). To do so, the main topics published by the international press on these efforts will be analysed. To this end, the following research questions are posed:

- RQ1: What has been the evolution of the presence of LVMH's sustainability and circular economy efforts in the international press?
- RQ2: What are the main contents that the international press has published about LVMH and its progress in sustainability and circular economy?
- RQ3: What are the most strategic themes picked up by the international press to link LVMH's fashion brands to SDGs 12 and 13?

Therefore, the objective of this study is to evaluate if LVMH gains visibility through information published on different media through analyzing information about sustainability, circular economy and climate change goals and solutions in this conglomerate of luxury brands. The contribution of this research is showing how circular economy is in the sustainability strategy of this luxury firm. It is a challenge and a business opportunity for the firm while responding to the need to reduce its environmental impact in the business strategy.

The chapter is organized as follows. First, a literature review has been conducted on sustainability in the luxury industry and CSR/Sustainability communication to better understand how these firms connect communication with the business strategy. Second, the methodology of this study is developed through quantitative and qualitative analysis to discover how media cover information on LVMH's sustainable activities. Third, we analyzed the results obtained for this case study providing specific results in some areas connecting with the sustainable strategy of this conglomerate. Finally, we include some conclusions and limitations of the study.

7.2 Sustainability in the Luxury Industry

In the last three decades, scholars have analyzed CSR considering the different responsibilities of companies and considering stakeholders' expectations from a voluntary perspective (Carroll, 2021). Furthermore, it is relevant to highlight the correlation between CSR and sustainability in companies for academic studies within a business perspective. Some scholars explain that CSR demonstrates how companies participate in contributing to a better society and the environment and it is a strategy linked to the culture; therefore, perceptions and expectations are key aspects for companies (Pedersen et al., 2018).

However, climate change is affecting business activities and sustainability is a strategy for companies from all sectors to fight global warming. From this perspective, sustainability is a long-term goal based on measuring environmental, societal and economic areas in companies to gain a competitive advantage through innovative and sustainable solutions to reduce the environmental impact (Thorisdottir & Johannsdottir, 2020). In other words, sustainability is a key driver of CSR and sustainable business models need to be aligned with the firm's culture and values to adopt sustainable practices and to engage different stakeholders (Pedersen et al., 2018). In sum, there is a need for companies to create a sustainable mindset based on actions and behaviours (Bertassini et al., 2021).

Together business and institutional regulations are working to reduce CO2 emissions. The Green Deal from the European Commission (EC), based on different commitments affecting European companies, such as circular economy, and non-financial reporting, among others, show that the EC is increasing innovative solutions to fight climate change, involving companies from all the industries to protect the planet, including the luxury industry (Rukhaya et al., 2021). This context also affects companies, considering the pressure for sustainability from stakeholders (Li & Leonas, 2019), especially the consumers' expectations demanding more transparency on sustainable business strategies (Campos et al., 2020).

Furthermore, some scholars argue that young consumers are demanding sustainable and environmental products and this pressure on luxury firms has also incentivised sustainable solutions for new products (Rukhaya et al., 2021). Consequently, the circular economy is leading to more sustainable business models (Brydges, 2021) and it is gaining interest (Olatubosun et al., 2021) to promote responsible consumption.

Because the circular economy is affecting the luxury industry, some researchers proposed in their study a conceptual framework to understand the challenges of the circular economy identifying the main drivers in the Textile and Apparel Industry supply chain, such as customer awareness of sustainable products, leadership and commitment from top managers, government funding and regulations pushing companies with guidelines and financial assistance (Jia et al., 2020).

Luxury companies are adopting sustainable business models and sustainable practices that include social practices to ensure good working conditions, protect the well-being of employees, and support social movements, among others; environmental practices (vertical integration in the production process, use of eco-friendly and vegan materials etc.); and economic practices (extending the product lifecycle, second-hand products) (Campos et al., 2020).

In this process, communication is also a crucial aspect of the business strategy and companies could engage stakeholders through communication in terms of sustainable communication to change their perceptions (Hepner et al., 2020). In fact, companies are including in their results reports on sustainability correlated with all the SDGs. The objective is to communicate social and environmental commitments to gain credibility through transparency (Campos et al., 2020; Gardetti, 2016).

7.3 CSR and Sustainability Communication

Communication is also relevant in the luxury industry to share values and sustainable practices with internal and external stakeholders, especially in a digital environment. In this area, the web becomes a key tool for interacting with all stakeholders (San Miguel et al., 2021) to transmit corporate statements and sustainable activities (Da Giau et al., 2016). The sustainability/CSR report, as an important document of the communication strategy, shows the results, impacts and shared value for society (San Miguel et al., 2021). Moreover, digitalization has impacted communication strategies with the highest visibility of company activities worldwide (Kotler et al., 2016). Consequently, it represents a risk and an opportunity for companies to manage perceptions on the Internet to build trust based on their CSR and sustainable activities (San Miguel et al., 2021) and to avoid cosmetic communication (Siano et al., 2016) and greenwashing (Szabo & Webster, 2021). As a result, communication is necessary to involve the whole company to act in a sustainable manner (Macgregor et al., 2020).

Some authors who have analyzed CSR communication in the luxury industry demonstrated that CSR messages that incorporated environmental benefits positively influenced consumers' perception from an environmental perspective, and that perception positively affected their evaluation of the firm (Kang & Sung, 2022).

To complete the strategic aspects affecting companies, communication is crucial for companies looking for a competitive advantage. CSR communication reflects in which specific areas the company is responsible for the environment and society. In general, websites include relevant information in this field (Mann et al., 2014) However, social media, Public Relations, and advertising campaigns, among others, are also strategic for the communication strategy of these companies (Abitbol et al., 2019; Troise & Camilleri, 2021).

Consequently, luxury firms have critical challenges affecting what and how these firms communicate in terms of their results and the company's sustainable value proposition to its stakeholders and especially with their customers in order to engage them (Schaltegger et al., 2016). Therefore, the CSR/Sustainability communication strategy should be understood in correlation with the CSR/Sustainability strategy of companies, particularly, in the luxury industry.

7.4 Methodology

LVMH has been chosen as a case study because it is the holding company of luxury brands that sells the most internationally (Deloitte, 2022). The LVMH group owns 14 fashion brands with an international presence in its Fashion & Leather Goods division, including Christian Dior, Kenzo, Loewe, Louis Vuitton, and Fendi, among others (LVMH, 2023), which represent 49% of the group's turnover (LVMH, 2022). In addition, LVMH (2022) has launched the LIFE 360 program, which encompasses the group's environmental strategy and focuses on circularity, biodiversity, the fight against climate change and traceability.

To analyze LVMH's presence in the international press when communicating its sustainability and circular economy efforts, a quantitative and qualitative analysis was used. The sample was obtained from the ProQuest tool, which was used to search for news items containing the terms LVMH, sustainability and circular economy. The period of analysis was between 2015 and 2023, given that the 2030 Agenda was adopted in 2015. After reviewing the news items obtained, those that were duplicated or whose content did not correspond to the object of study were discarded. In addition, the results provided by ProQuest were from 2017, so the final sample consisted of a total of 106 news items from international media dating between 2017 and 13 October 2023.

The quantitative analysis study considered variables such as the evolution of the number of news items collected by the media, the country, the type of media and a rootedness analysis carried out with ATLAS.ti 23.3.2 to determine the number of quotes that appear in the news items in the sample and that are associated with a code. The qualitative analysis was based on a content study (Patton, 1990) of the 106 news items also using the ATLAS.ti 23.3.2 tool. A two-step inductive coding was performed: first, a coding was done based on the results provided by the artificial intelligence (AI) tool available in the version of ATLAS.ti 23.3.2; and secondly, more than 2000 codes were cleaned to select those closest to the object of study and those related to SDGs 12 (responsible consumption and production) and 13 (Climate action). In total, 18 codes were obtained (Table 7.1) which were divided by categories (Bardin, 1996) into three blocks distinguishing between the most general (CG), the specific (SC) and the associated (AC), so that the relationship between CG and SC, together with CG and AC, could be studied through the concurrence analysis allowed by ATLAS.ti (Llorente-Barroso et al., 2022).

The study of co-occurrences (ATLAS.ti, 2023) makes it possible to relate the confluences or synergies between codes based on the number of times (Gr) that a quotation associated with a code appears in each news item. In this way, the strength

Table 7.1 Codebook or code tree for qualitative analysis

General codes (GC)	Specific codes (SC)	Associated codes (CA)
Code	Code	Code
Sustainability	Responsible consumption and production	Natural resources
Circular economy	Environment protection	Raw materials
Luxury fashion industry	Climate change	Waste and chemical management
Ethics	Energy management	CO2 emissions
	CSR	Regulations
	Global trends	Certifications
	Retail transformation	Brand management

Source Authors

or intensity between ideas will be determined, the result of which will be shown in a table. Based on the results of the co-occurrence study, two types of graphs will be used to reinforce the explanation of the connection between codes or ideas (ATLAS.ti, 2023). Firstly, the width of the lines connecting the different terms on either side of the Sankey diagram represents the relationship between the codes. Secondly, the Strength chart will help to visualize the dominant codes in the relationships through the proximity between ideas according to the size at which the codes are represented circularly, as well as by the closeness or remoteness of each sphere.

7.5 Results

7.5.1 Quantitative Analysis

The main results of the quantitative analysis focus on the numerical study of the number of news items according to three variables: evolution by year, by type of article and by country of origin.

As can be seen from the number of articles per year (Graph 7.1) and considering that the data for 2023 does not cover the whole year (up to 13 October), the press coverage of LVMH's sustainability efforts is increasing, thus showing an upward trend with peaks in 2019 and 2022. It should be noted that the 2020 results correspond to the consequences of the international COVID-19 pandemic.

Regarding the typology of media articles (Graph 7.2), ProQuest has categorised the results into four types: newspapers, magazines, industry media and blogs, podcasts or websites. Of these, industry media published the most stories on LVMH's sustainability efforts with 58 stories and newspapers with 29 stories.

In the analysis of the number of news items by country (Graph 7.3), the highest volume of news items published about LVMH in the U.S. stands out, with 61 news

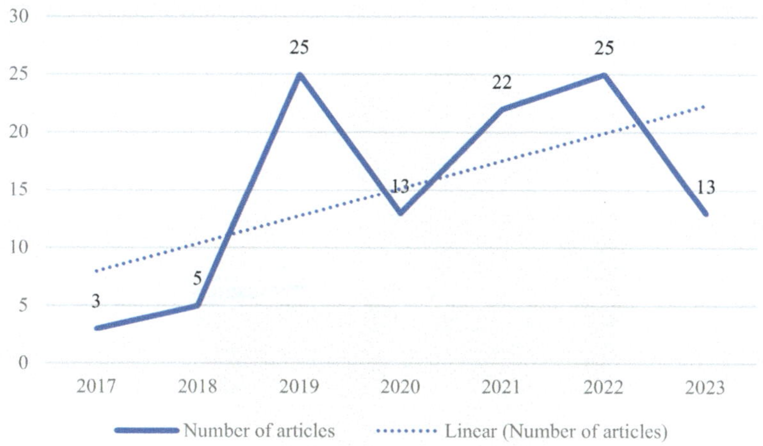

Graph 7.1 Number of articles per year. *Source* Authors based on ProQuest data

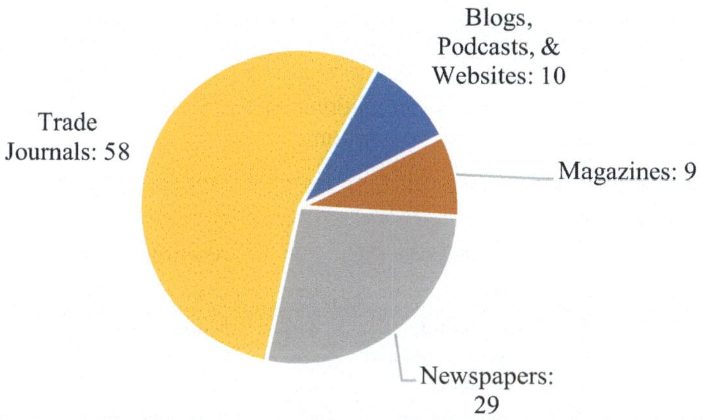

Graph 7.2 Number of articles per type. *Source* Authors based on ProQuest data

items in the six years studied. While it is true that this data could be biased by the tool's capacity to collect data by number of media and language, it demonstrates the interest of the most influential international press in publishing the LVMH Group's efforts in sustainability. Finally, to contextualise the results of the subsequent qualitative analysis, the number of times (Gr) that a quote appears associated with a given code is presented to represent the frequency with which this code appears in the sample analysed.

In the rootedness analysis (Table 7.2) we can see the number of citations (Gr) that have been coded for each code. There is a higher rootedness in the codes 'Sustainability' with 862 citations, 'Environment protection' with 416 citations and 'Responsible consumption and Production' with 416 citations. In contrast, the codes with the

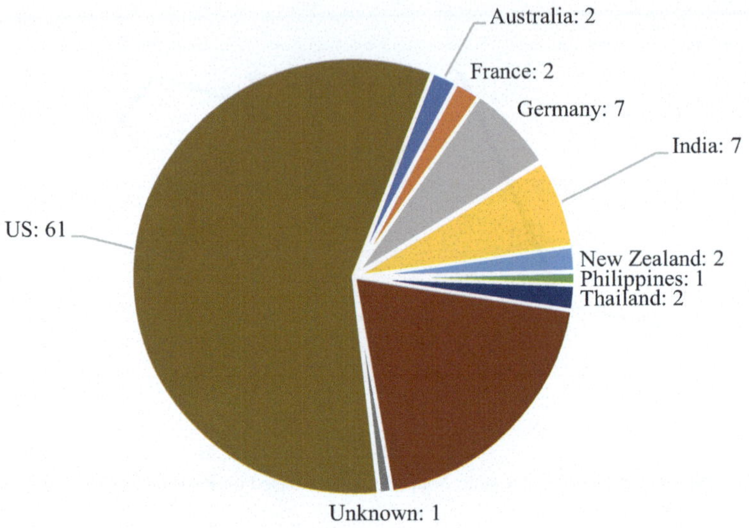

Graph 7.3 Number of articles per country. *Source* Authors based on ProQuest data

lowest degree of rootedness are 'Certifications' with 18 citations, 'Ethics' with 39 citations and 'Raw materials' with 43 citations.

Table 7.2 Rooting of the codes

General codes (GC)		Specific codes (SC)		Associated codes (CA)	
Code	Gr	Code	Gr	Code	Gr
Sustainability	862	Responsible consumption and production	416	Natural resources	72
Circular economy	142	Environment protection	575	Raw materials	43
Luxury fashion industry	265	Climate change	125	Waste and chemical management	66
Ethics	39	Energy management	103	Carbon footprint	61
		CSR	225	Regulations	56
		Global trends	114	Certifications	18
		Retail transformation	101	Brand management	343

Source Own elaboration with ATLAS.ti 23.3.2

7.5.2 *Qualitative Analysis*

The results of the qualitative content analysis based on the analysis of co-occurrences between general codes (GC) and specific codes (SC), and between general codes (GC) and associated codes (AC) are presented below, followed by a Sankey diagram and a Strength chart.

According to the first co-occurrence analysis of the general codes (GC) associated with the object of study and those more specific (SC) (Table 7.3), there is a very high degree of connection between 'Sustainability' and 'Environmental protection' with 239 co-occurrences and between 'Sustainability' and 'Responsible consumption and production' with 154 co-occurrences. These specific codes are linked respectively to SDGs 13 (Climate action) and SDG 12 (Responsible consumption and production).

On the other hand, it is striking that the degree of co-occurrence generated by the code 'Ethics' with the rest of the topics studied is very low, almost non-existent, which suggests that it is content that arouses little interest among the international press, either on its own or linked to other topics.

The Sankey diagram above (Fig. 7.1) visually summarises the contribution of each general and specific code to the construction of the 'Luxury fashion industry' narrative. As can be seen, there is a strong connection with 'Sustainability' in relating concepts to reflect the efforts made by the fashion brands of the LVMH

Table 7.3 Co-occurrence analysis of general codes (GC) and specific codes (SC)

	Circular economy Gr = 142	Ethics Gr = 39	Luxury fashion industry Gr = 265	Sustainability Gr = 862
Climate change Gr = 125	9	0	5	49
CSR Gr = 225	6	6	6	64
Energy management Gr = 103	3	0	0	20
Environmental protection Gr = 575	41	6	26	239
Global trends Gr = 114	0	1	15	8
Responsible consumption and production Gr = 416	23	4	22	154
Retail transformation Gr = 101	1	0	4	10

Source Authors with ATLAS.ti 23.3.2

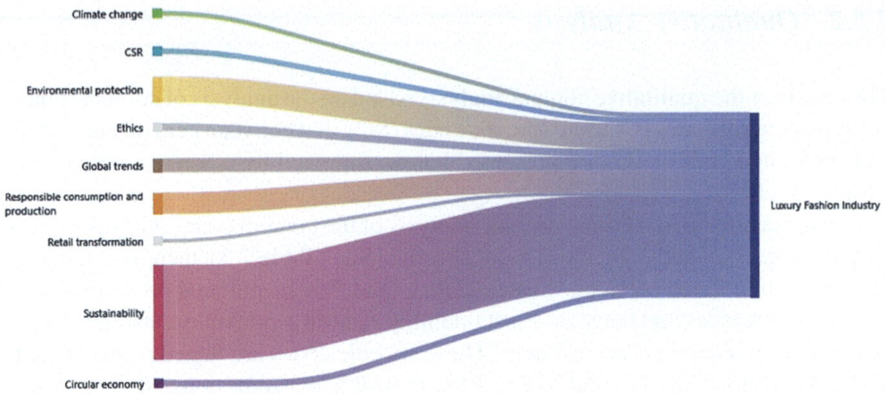

Fig. 7.1 Connection of 'Luxury fashion industry' with general codes (GC) and specific codes (SC). *Source* Authors with ATLAS.ti 23.3.2

group. Secondly, it can be seen how 'Luxury fashion industry' is also related to 'Environmental protection' and 'Responsible consumption and production'.

Taking the explanation of the results a step further, the Strength diagram (Fig. 7.2) represents the intensity of the relationship of the codes GC and SC with 'Luxury fashion industry'. Beyond what has been explained so far, this figure stands out because it shows a second relationship of 'Luxury fashion industry' themes with 'CSR', 'Circular economy' and 'Climate change'. Furthermore, it is striking that the 'Energy management' code is not linked to the 'Luxury fashion industry' in terms of the line that connects them.

The second co-occurrence analysis corresponds to the study of general codes (GC) and associated codes (AC) (Table 7.4) and is in line with the previous study (Table 7.3). 'Sustainability' is the code with the most co-occurrences with the rest of the associated codes (AC). Firstly, the high connection of 'Sustainability' with 'Carbon footprint' stands out with 32 co-occurrences and its linkage with SDG 13 (Climate action). In addition, 'Sustainability' is related to 'Waste and chemicals management', thanks to the 27 co-occurrences that have been established, and therefore its connection with SDG 12. On the other hand, 'Sustainability' and 'Brand management' are also linked with 26 co-occurrences where a first connection with the management of luxury fashion brands can be observed.

Finally, as in the first co-occurrence analysis, there is a very low connection of the term 'Ethics' with the rest of the associated codes studied. This could show that although it is a general term linked to the theoretical framework, the press hardly uses this term to elaborate the content of its news pieces.

The second Sankey diagram (Fig. 7.3) reinforces the high linkage illustrated above of 'Luxury fashion industry' with 'Sustainability'. Given the width of the line blocks, the linkage of 'Luxury fashion industry' with 'Brand management' is also noteworthy. In addition, the linkage of other sub-themes such as 'Raw materials', 'Ethics', 'Circular economy', and 'Carbon footprint' is also noteworthy.

Fig. 7.2 Relationship of 'Luxury fashion industry' strength with the general codes (GC) and the specific codes (SC). *Source* Authors with ATLAS.ti 23.3.2

Table 7.4 Co-occurrence analysis of general codes (GC) and associated codes (AC)

	Circular economy Gr = 142	Ethics Gr = 39	Luxury fashion industry Gr = 265	Sustainability Gr = 862
Brand management Gr = 343	8	3	20	26
Carbon footprint Gr = 61	5	0	6	32
Certifications Gr = 18	3	1	0	10
Natural resources Gr = 72	3	1	3	12
Raw materials Gr = 43	2	0	8	16
Regulations Gr = 56	4	1	2	8
Waste and chemicals management Gr = 66	4	0	3	27

Source Authors with ATLAS.ti 23.3.2

Fig. 7.3 Connection of 'Luxury fashion industry' with general codes (GC) and associated codes (AC). *Source* Authors with ATLAS.ti 23.3.2

The Strength chart between 'Luxury fashion industry' with the general codes (GC) and the associated codes (AC) (Fig. 7.4) confirms what was shown in the Sankey diagram in terms of the relationship with 'Sustainability' and 'Brand management'. However, the picture orders a third link of importance with 'Circular economy' which was not so clear in Fig. 7.3. Furthermore, it is worth noting the more distant connection of 'Regulations' with 'Luxury fashion industry' and a decoupling from the code 'Certifications'.

To conclude the presentation of the results, a summary of the main findings is presented below:

- There is an upward trend in international press coverage of LVMH's sustainability and circular economy efforts.
- The trade media are the ones that publish the most news about the object of study of this research.
- By country, the United States accounts for the largest number of news items published about LVMH.
- 'Sustainability' is the code with the highest number of citations from LVMH's news gathering, and therefore, the most used by the international press. At the other extreme is the code related to 'Certifications'.
- In LVMH coded news, 'Luxury fashion industry' has a high degree of connection with the codes 'Sustainability', 'Environmental protection' and 'Responsible consumption and production'. On the other hand, there is a low linkage with the codes 'Energy management', 'Ethics' and 'Certifications'.

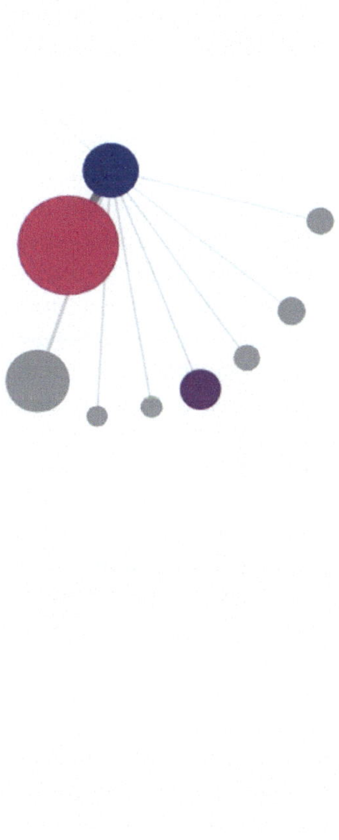

Fig. 7.4 Strength relationship of 'luxury fashion industry' with general codes (GC) and associated codes (AC). *Source* Authors with ATLAS.ti 23.3.2

7.6 Discussion and Conclusions

Sustainability in luxury companies has been a priority in recent years, because of CO2 emissions generated by this industry, especially the fashion industry. On the one hand, this study shows the sustainable commitment of a leading luxury brand in the market, as well as the sustainable and innovative solutions it includes in its production process (Thorisdottir & Johannsdottir, 2020). All of this is evaluated through the information available about the company in different communication platforms (San Miguel et al., 2021; Troise & Camilleri, 2021) and especially through the media. It can be noted that the firm has visibility in the media from a sustainable perspective, as shown in this analysis. Therefore, this information should be analysed together with the company's communication strategy to gain transparency and credibility (San Miguel et al., 2021). All of this is because it has an impact on the global perceptions of the company's stakeholders, especially consumers (Pedersen et al., 2018).

Sustainability and CSR are increasing in the global market in all sectors since global demands and risks place sustainability as a priority and competitive advantage for companies. Furthermore, measuring and communicating is necessary for companies to allow the business contribution to sustainable development to be evaluated. In this sense, the proposals of the UN (SDGs) and EC (The Green Deal) are global initiatives that invite companies around the world to report measurable information to specifically evaluate business contributions.

Regarding the evolution of the presence of LVMH's sustainability and circular economy efforts in the international press, it is noteworthy that sustainability has become strategic content for the media as it has experienced an increasing trend since 2017. The media that mostly cover this type of news are the trade media, which is understandable given that they have more space to cover a greater number of retail news stories. By country, despite the possible biases of the tool, the United States represents the largest coverage of LVMH's sustainability efforts. One reason may be that LVMH's Leather & Goods division in the United States represents the group's second-largest market (LVMH, 2022).

The main content published in the international press about LVMH and its progress in sustainability and the circular economy is based on the thematic axis of sustainability to explain the efforts made by the fashion industry. On a second level, there is content related to environmental protection and sustainable consumption and production. In fact, this link shows that the international press does pick up on the efforts made by LVMH fashion brands around SDGs 12 and 13, so it can be concluded that communicating sustainability efforts is relevant for the international press and therefore a strategic area to be promoted by fashion companies.

In addition, a second level of strategic themes that the international press highlights about LVMH is striking. On the one hand, one could distinguish a theme related to strategic business management such as brand management, CSR, and corporate ethics. And on the other hand, a set of themes that are again related to SDGs 12 and 13, the circular economy, climate change and carbon footprint. In this sense, it could be argued that LVMH establishes a link between strategic and operational issues when communicating sustainability externally. Indeed, given that one of LVMH's values (2022) is its commitment to making a positive impact and creating a better society, one of its strategic areas of action, it is likely that LVMH is including reputational impact in its performance metrics that can be derived from its sustainability efforts when reaching out to its most influential stakeholders in this regard. Therefore, this analysis serves to answer the questions RQ1, RQ2 and RQ3 in this case study with these specific results with the quantitative and qualitative analysis.

The implications of this study for policymakers and practitioners are clear. First, this research provides arguments to align the sustainability strategy with the communication strategy showing the company's main activities in this area. Second, the information they publish (press media, etc.) can influence the perception and behaviors of consumers when they promote sustainable consumption and circular economy to engage them with responsible practices. Third, communication also allows managers to show the real contribution of companies to sustainable development, increasing their transparency and reputation when they communicate their level of progress even on the 17 SDGs. Finally, this information provides the opportunity to share sustainable practices for different industries and create alliances between them.

However, this study has some limitations. First, the authors only include some areas of the company's sustainable strategy related to SDG 12 and SDG 13. In fact, the company includes activities connecting with the 17 SDGs in the annual Sustainability Report; but this study is only focused on analysing the information related to fighting climate change and its connection with circular economy and responsible consumption. Second, the authors only analyse one industry, the luxury industry, influenced by the need to reduce its CO_2 emissions in their production compared with other sectors. Third, the study only includes the visibility of LVMH in press media worldwide following this quantitative and qualitative methodology. Therefore, the conclusions are only significant to know the presence of this firm in the media, but the results can inspire scholars to use this methodology to compare results with other firms from different industries.

Finally, the results of the research can serve as an example for other fashion brand companies to enhance their communication on sustainability-related issues, without the fear of falling into greenwashing that is not based on tangible facts. In fact, this study highlights LVMH's direct linking of sustainability to SDGs 12 and 13 to justify its efforts and concrete measures in its fashion brand division. Therefore, future research could compare the efforts of other international luxury fashion brand holding companies such as Kering or Chanel to learn more and contribute to better communication of sustainability and the SDGs.

References

Abitbol, A., Meeks, J., & Cummins, R. G. (2019). Does oil and goodwill mix? Examining the oil and gas industry's impact on stakeholder engagement on Facebook. *Environmental Communication, 13*(2), 192–208. https://doi.org/10.1080/17524032.2018.1546751

ATLAS.ti. (2023). ATLAS.ti 23 Windows. https://doc.atlasti.com/QuicktourWin.es/ATLAS.ti_QuickTour_es_Win.23.pdf?_ga=2.150227254.1860783379.1698000280-1675242112.1698000280

Bardin, L. (1996). *Análisis de contenido*. Akal

Bertassini, A. C., Ometto, A. R., Severengiz, S., & Gerolamo, M. C. (2021). Circular economy and sustainability: The role of organisational behaviour in the transition journey. *Business Strategy and the Environment, 30*(7), 3160–3193.

Brydges, T. (2021). Closing the loop on take, make, waste: Investigating circular economy practices in the Swedish fashion industry. *Journal of Cleaner Production, 293*, 126245.

Campos, J., Hussain, D., & McColl, R. (2020). Luxury fashion and sustainability: Looking good together. *Journal of Business Strategy, 41*(4), 55–61.

Carroll, A. B. (2021). Corporate social responsibility: Perspectives on the CSR construct's development and future. *Business & Society*. https://doi.org/10.1177/

Da Giau, A., Macchion, L., Caniato, F., Caridi, M., Danese, P., Rinaldi, R., & Vinelli, A. (2016). Sustainability practices and web-based communication: An analysis of the Italian fashion industry. *Journal of Fashion Marketing and Management, 20*(1), 72–88.

Deloitte. (2022). Global Powers of Luxury Goods 2022. https://www.deloitte.com/content/dam/assets-shared/legacy/docs/analysis/2022/gx-global-powers-of-luxury-goods-report.pdf

Gardetti, M. A. (2016). Loewe: Luxury and sustainable management. *Handbook of Sustainable Luxury Textiles and Fashion:, 2*, 1–16.

Hepner, J., Chandon, J. L., & Bakardzhieva, D. (2020). Competitive advantage from marketing the SDGs: A luxury perspective. *Marketing Intelligence & Planning, 39*(2), 284–299. https://doi.org/10.1108/MIP-07-2018-0298

Jia, F., Yin, S., Chen, L., & Chen, X. (2020). The circular economy in the textile and apparel industry: A systematic literature review. *Journal of Cleaner Production, 259*, 120728.

Kang, E. Y., & Sung, Y. H. (2022). Luxury and sustainability: The role of message appeals and objectivity on luxury brands' green corporate social responsibility. *Journal of Marketing Communications, 28*(3), 291–312.

Kotler, P., Kartajaya, H., & Setiawan, I. (2016). *Marketing 4.0: Moving from traditional to digital*. John Wiley & Sons.

Kunz, J., May, S., & Schmidt, H. J. (2020). Sustainable luxury: Current status and perspectives for future research. *Business Research, 13*, 541–601.

Li, J., & Leonas, K. K. (2019). Trends of sustainable development among luxury industry. In *Sustainable Luxury* (pp. 107–126). Springer. https://doi.org/10.1007/978-981-13- 0623–5_6.

Llorente-Barroso, C., Mena-Moreno, J., Ferreira, I., & García-García, F. (2022). El compromiso social de las grandes empresas con la discapacidad en España y Portugal: Análisis de las iniciativas de empresas del IBEX35 y del PSI20. *Revista Prisma Social, 36*, 4–36. https://revistaprismasocial.es/article/view/4567

LVMH. (2022). LVMH Snapshot–2022 Figures. https://r.lvmh-static.com/uploads/2023/01/lvmh_snapshot_2022_en.pdf

LVMH. (2023). Fashion & Leather Goods. https://www.lvmh.com/houses/fashion-leather-goods/

Macgregor, R. K., Sroka, W., & Pelikánová, R. M. (2020). The CSR perception of front-line employees of luxury fashion businesses. *Organisacija, 53*(3), 198–211.

Mann, M., Byun, S. E., Kim, H., & Hoggle, K. (2014). Assessment of leading apparel specialty retailers' CSR practices as communicated on corporate websites: Problems and opportunities. *Journal of Business Ethics, 122*, 599–622.

Olatubosun, P., Charles, E., & Omoyele, T. (2021). Rethinking luxury brands and sustainable fashion business models in a risk society. *Journal of Design, Business & Society, 7*(1), 49–81. https://doi.org/10.1386/dbs_00020_1

Osburg, V. S., Davies, I., Yoganathan, V., & McLeay, F. (2021). Perspectives, opportunities and tensions in ethical and sustainable luxury: Introduction to the thematic symposium. *Journal of Business Ethics, 169*(2), 201–210. https://doi.org/10.1007/s10551-020-04487-4

Ozdamar-Ertekin, Z. (2019). Can luxury fashion provide a roadmap for sustainability? Markets, *Globalization & Development Review, 4*(1). https://doi.org/10.23860/MGDR2019-04-01-03

Pedersen, E. R. G., Gwozdz, W., & Hvass, K. K. (2018). Exploring the relationship between business model innovation, corporate sustainability, and organisational values within the fashion industry. *Journal of Business Ethics, 149*, 267–284.

Pencarelli, T., Gabbianelli, L., & Savelli, E. (2020). The tourist experience in the digital era: The case of Italian millennials. *Sinergie Italian Journal of Management, 38*(3), 165–190.

Patton, M. Q. (1990). *Qualitative evaluation and research methods*. Sage.

Rukhaya, S., Yadav, S., Rose, N. M., Grover, A., & Bisht, D. (2021). Sustainable approach to counter the environmental impact of fast fashion. *The Pharma Innovation Journal, 10*(8S), 517–523.

San Miguel, P., Pérez-Bou, S., Sádaba, T., & Mir-Bernal, P. (2021). How to communicate sustainability: From the corporate Web to E-commerce. The case of the fashion industry. *Sustainability, 13*(20), 11363.

Schaltegger, S., Hansen, E. G., & Lüdeke-Freund, F. (2016). Business models for sustainability: Origins, present research, and future avenues. *Organisation & Environment, 29*(1), 3–10.

Siano, A., Conte, F., Amabile, S., Vollero, A., & Piciocchi, P. (2016). Communicating sustainability: An operational model for evaluating corporate websites. *Sustainability, 8*(9), 950.

Szabo, S., & Webster, J. (2021). Perceived greenwashing: The effects of green marketing on environmental and product perceptions. *Journal of Business Ethics, 171*, 719–739.
Thorisdottir, T. S., & Johannsdottir, L. (2020). Corporate social responsibility influencing sustainability within the fashion industry. A systematic review. *Sustainability, 12*(21), 9167.
Troise, C., & Camilleri, M. A. (2021). The use of digital media for marketing, CSR communication and stakeholder engagement. In *Strategic corporate communication in the digital age* (pp. 161–174). Emerald Publishing Limited.
Willersdorf, S., Hazan, J., Ricci, G., Prénaud, A., Bianchi, F., Seara, J., & Yang, V. (2020). A new era and a new look for luxury. *BCG Global.*

Correction to: Textile Industry and Sustainability: An Integrated Examination

Cristina Blanco González-Tejero, Samuel Ribeiro-Navarrete, Raquel Diaz, and Ruben Furio-Piqueres

Correction to:
Chapter 1 in: J. F. Gallego-Nicholls and A. Carrilero-Castillo (eds.), *Textile Industry and the SDGs*, SDGs and Textiles, https://doi.org/10.1007/978-981-97-5073-3_1

The original version of the chapter was inadvertently published with the incorrect author list, which has been updated as follows: Blanco-González-Tejero to Cristina Blanco González-Tejero, Samuel Ribeiro-Navarrete, Raquel Diaz, Ruben Furio-Piqueres. The chapter has been updated with the changes.

The updated version of this chapter can be found at
https://doi.org/10.1007/978-981-97-5073-3_1

J. F. Gallego-Nicholls and A. Carrilero-Castillo (eds.), *Textile Industry and the SDGs*, SDGs and Textiles, https://doi.org/10.1007/978-981-97-5073-3_8

Correction to: Textile Industry and Sustainability: An Integrated Examination

Correction to:

SPRINGER NATURE

GPSR Compliance

The European Union's (EU) General Product Safety Regulation (GPSR) is a set of rules that requires consumer products to be safe and our obligations to ensure this.

If you have any concerns about our products, you can contact us on ProductSafety@springernature.com

In case Publisher is established outside the EU, the EU authorized representative is:

Springer Nature Customer Service Center GmbH
Europaplatz 3
69115 Heidelberg, Germany

The manufacturer's authorised representative in the EU is Springer Nature Customer Service Centre GmbH, Europaplatz 3, 69115 Heidelberg, Germany. If you have any concerns regarding our products, please contact ProductSafety@springernature.com

Printed and bound by CPI Group (UK) Ltd, Croydon, CR0 4YY
01/10/2026
02225944-0007